Science and Mysticism

Science and Mysticism

A Comparative Study of Western Natural Science, Theravāda Buddhism, and Advaita Vedānta

Richard H. Jones

Lewisburg
Bucknell University Press
London and Toronto: Associated University Presses

Associated University Presses
440 Forsgate Drive
Cranbury, NJ 08512

Associated University Presses
25 Sicilian Avenue
London WC1A 2QH, England

Associated University Presses
2133 Royal Windsor Drive
Unit 1
Mississauga, Ontario
Canada L5J 1K5

The paper used in this publication meets the requirements of the American National Standard for Permanence of Paper for Printed Library Materials Z39.48-1984.

Library of Congress Cataloging-in-Publication Data

Jones, Richard H., 1951–
Science and mysticism.

Bibliography: p.
Includes index.
1. Religion and science—1946– . 2. Mysticism—India. 3. Buddhism and science. 4. Hinduism and science. I. Title.
BL240.2.J66 1986 291.1'75 84-46098
ISBN 0-8387-5093-1 (alk. paper)

Printed in the United States of America

To my Father, Curtis Knowles Jones

Contents

Abbreviations 9
Introduction 11

PART I: *The Frameworks of Science and Mysticism*

1 Science 19
2 Mysticism 41

PART II: *The Nature of Scientific and Mystical Claims Compared*

3 Ways of Life and Knowledge-Claims 73
4 Reality 81
5 The Nature of Knowledge 110
6 Experiences 127
7 Language 140

PART III: *Specific Scientific and Mystical Claims Compared*

8 Possible Relationships between Scientific and Mystical Claims 169
9 Cosmology, Physics, and Mysticism 180

PART IV: *A Reconciliation of Science and Mysticism*

10 A Reconciliation of Science and Mysticism 211

Appendix: Concerning the Possible Philosophical Significance of Scientific Studies of Mysticism 219
Notes 234
Select Bibliography 254
Index 263

Abbreviations

For Theravāda Buddhism:

D	*Dīgha-nikāya*
M	*Majjhima-nikāya*
A	*Aṅguttara-nikāya*
S	*Saṁyutta-nikāya*
Dhp	*Dhammapada*
Mlp	*Milindapañha*

Numbers refer to the Pāli Text Society editions

For Advaita Vedānta:

BS(B)	*Brahma-sūtra (Bhāṣya)*
BG(B)	*Bhagavad-gītā (Bhāṣya)*
BU(B)	*Bṛhadāraṇyaka Upaniṣad (Bhāṣya)*
CU(B)	*Chāndogya Upaniṣad (Bhāṣya)*
KU(B)	*Kena Upaniṣad (Bhāṣya)*
MU(B)	*Muṇḍaka Upaniṣad (Bhāṣya)*
TU(B)	*Taittirīya Upaniṣad (Bhāṣya)*

Introduction

The Issues

In the past few years, popular books relating scientific thought and various Eastern religious traditions have proliferated. Commentators such as Fritjof Capra in his *The Tao of Physics* and Gary Zukav in his *The Dancing Wu Li Masters* have claimed that contemporary physics has returned to an ancient mystical view of reality. To get a critical edge on the problems involved in such a comparative undertaking, what will be examined here are the claims about reality in Western natural science and in two Indian traditions. Are the mystical claims of the Asian traditions of the same nature as the scientific ones? Do mystics ask the same types of questions as scientists? Do they provide different answers? Or do they answer different questions? Do their answers entail claims that contradict or agree with those of the scientists? If not, what are the possible relationships between their claims? Such philosophical issues will guide this project.

Methodological Considerations

As a preliminary step, some regard must be paid to this project's methodology. Since one cannot reasonably be expected to answer every philosophical question involved in any inquiry, certain assumptions must be made. The best that can be done is to state the more problematic of one's own assumptions. For example, one such assumption is that contemporary post-empiricist philosophers of science, who are concerned more with the actual history of science, reflect better the nature of the scientific enterprise than do those philosophers concerned with a formal reconstruction of scientific thought.

The traditions to be employed here as representative of mysticism are the Theravāda tradition of Buddhism and Śaṁkara's Advaita Vedānta. These have been chosen because writers on the subject of science and Eastern thought look upon these, along with Taoism and certain Mahāyāna Buddhist traditions, as the substance of "Eastern thought."[1] Problems with characterizing all of Asian thought as being of one fabric can be brought out by looking at just these two traditions. Classical texts as they stand today will be the sole source of information concerning both: for the Theravāda tradition, the Pāli *Nikāyas* and the *Milindapañha;* and for Advaita, Śaṁkara's commentaries on the *Brahma-sūtras* and the major *Upaniṣads.*

A key assumption for this comparative study is that approaching mysticism through discourses intended to be instructions and explanations for the unenlightened will supply sufficient understanding. That is, a certain "mystical sensitivity" may be necessary to follow these analogies and descriptions, but it is assumed that a mystical experience or enlightenment is not necessary for all understanding in this field.[2] Certainly such texts as the *Milindapañha* and *Brahma-sūtra-bhāṣya* are books of arguments and responses, with many of the objections and replies sounding very "Western" in character—they are not so difficult to understand as the label "mystical" might suggest.

With regard to the mystical traditions, the focus of attention will be the belief-content of the religious doctrines and practices. A "belief" is the propositional content of what is argued for or taken for granted by specific actions and claims. Many background beliefs are not explicitly formulated (e.g., the Buddha does not provide any extended justification of *karma* or rebirth) because these are more implicitly presupposed than points of contention. Believers do not speak of what they believe but of what, from their point of view, they *know* to be true of reality and themselves. Only when challenged do believers have occasion to question their tenets (which requires attempting to make them explicit). Belief-commitments that can be abstracted from the Theravāda claim that the enlightened Buddha gained release from the cycle of rebirths are: (1) There is a cycle of rebirths; (2) It is possible to terminate the cycle; and (3) Someone has done so. Such beliefs are clearly not the total substance of a full religious faith, but they are not merely one expression of it either; without the acceptance of the belief-claims once they are presented, what "adherence to religious faith" means is not obvious. Thus belief-claims form a legitimate, if not absolutely vital, topic of study. Believers are committed to specific belief-claims even if their particular culture has no concept of "belief"; it is no less legitimate for reflecting an observer's set of interests rather than the participants'.

The objectives of this project will be achieved by examining together the presuppositional beliefs of two *prima facie* differing enterprises (science and mystical ways of life) in two differing cultures (the West and India). The common ground needed to compare science and the two Indian traditions is that each embodies conceptual schemes intending knowledge of reality. Within this framework, differences and similarities of concerns, the epistemological placing of different experiences, visions of the world, specific claims about the world, and so forth can be fruitfully studied. Operating outside mystical awareness to compare mysticism and science does not beg any question against mysticism since the importance of such experiences and values can still be accepted—in fact, mystics themselves may conceptualize and evaluate their experiences only outside the mystical depth, including ordinary nonmystical states of awareness. An ever-present danger is letting one of the compared subjects determine the whole project—for example, letting science shape our view of the Indian traditions and then employing this truncated version of the latter for the purpose of comparison to science. The chief risk is coming up with descriptions that make sense to us but completely miss the original content. Our understanding in general, including science, will come to bear on our understanding of Indian religiosity. Thus the *religious* contexts of the Indian way of life will need to be repeatedly emphasized. But only those aspects of each enterprise which become prominent when science and mysticism are investigated together are ultimately important for this project.

The Potential Value of Comparing Science and Mysticism

> Everyone knows how useful it is to be useful.
> No one seems to know how useful it is to be useless.
> —Chuang tzu

This project is a work in metaphysics in the sense of encompassing diverse areas of human experience. Such a philosophical interest relies totally upon other enterprises, but it has a place of its own: being a great scientist or mystic does not mean that one knows the problems related to interpreting the status of one's insight or the problems involved in making comparisons between different enterprises. The justification for comparative studies is that they often bring to light aspects and presuppositions of both compared subjects that are not apparent when the focus of attention is restricted to more limited topics within only one subject. For instance, the subject-object stance of science is a tacit presupposition whose significance can be brought out only by contrast with another approach to the world. For mysticism, the relation of concepts to experiences in science may prove helpful for understanding the roles of background cultural factors and experiential factors in the discovery and justification of each mystical system.

Werner Heisenberg once remarked that generally "in the history of human thinking the most fruitful developments frequently take place at those points where two different lines of thought meet."[3] For scientists, mystical texts may be sources of insights for the development of scientific theories—better sources, in fact, than many nonscientific materials, since mystical systems also attempt to understand the world. Mystics may develop new metaphysical systems using scientific theories as root-metaphors or as sources of explanatory images; or they may gain awareness of their own presuppositions, thereby facilitating an end to attachment. Or perhaps some entirely other development may occur. It can be reasonably hoped, however, that for the rest of us this project may contribute to a more informed view of each enterprise and of the possible relations between them.

There is no simple conclusion to state in advance. It will be seen that mysticism and science converge and diverge in a variety of complex ways, ways not always detrimental to mysticism and favorable to science. The procedure will be (1) to give the general frameworks of science, Theravāda Buddhism, and Advaita Vedānta as knowledge-intending enterprises, on the basis of which (2) to compare and contrast the nature of the claims in each, and (3) to compare and contrast some of the claims themselves. A reconciliation of science and mysticism is also proposed; an appendix in which the philosophical significance of scientific studies of meditation is discussed is also included.

Science and Mysticism

PART I

The Frameworks of Science and Mysticism

1
Science

Understanding Nature

Albert Einstein long remembered the wonder that was touched off in him when his father first showed him a compass: "Something deeply hidden had to be behind things."[1] To grasp such unobserved processes is to *understand* reality to the extent that understanding is possible within a scientific framework. Scientists develop conceptual schemes that explain the flow of events by means of these real workings beneath appearances. The world is thereby rendered intelligible insofar as what we expect to occur does in fact occur. The quest to understand has been more central to modern science than the practical side of science, that of the effort to control nature. Hence two of the most dramatic scientific revolutions—the Copernican and Darwinian—have encountered great social hostility without having any direct influence on most people's material needs. The Baconian notion of science as "power-thought" has also been present in science, reflecting our need to know the relation between events in order to avoid the dangerous and to cultivate the advantageous. Today the distinction between "science" and "technology" is made more difficult with the mutual penetration of science and industry.[2] Nevertheless, as Stephen Toulmin has pointed out, "through most of its three-hundred-year history modern science has been associated far more closely with theology than with technology."[3] The sciences, in other words, are more "natural philosophy" and "natural history" than industry.

Concepts

In a very important sense, the sciences operate in a simpler, artificial world of human constructions rather than in reality full blown. Scientists begin with everyday perceptions that abstract points of relative stability from the total flux of experience. The continuous stream of experience is thereby replaced in our awareness by reality filtered through distinct concepts ordering experiences in terms of convenient categories that abstract rough similarities between unique items. In the everyday world, posits such as tables and chairs organize our experiences; they provide constants throughout the conflicting experiences that result from changing angles of visions and so forth. Scientists relate and revise these concepts into systems that are overall simple and fruitful in explaining our experiences. Everyday concepts are more firmly embedded within our total picture of the world, but even they are open to revision in light of new ideas and experiences. More theoretical concepts, posited for understanding more abstract patterns behind surface appearances, change more frequently as theories evolve because the evidence for them is less direct: inferring the existence of theoretical entities is comparable to inferring the existence of a chair from marks on a rug interpreted as tracks caused by dragging the legs of the chair across it. The necessity of postulating an entity behind the experience is still there, in addition to complex relations to direct evidence. In each case the posits may be real, or we may be mistaken in our understanding of the processes of reality.

In this way, scientists search for invariants in the flux of experience. Their concepts are attempts at isolating central relations between occurrences, i.e., the *structures* of reality. But the importance of motion and time as variables in physics ought to suffice to show that the sciences do not view reality as static. Even "mass" can only be measured in terms of such interactions as resistance to changes in motion. What always remains central are the interactions of entities and their relation to experienced reality, not the entities in isolation.

Observation and Objectivity

Concepts thus figure prominently in the only type of experience recognized as cognitive within a scientific framework, i.e., observation. Initially, sense-experience in science is construed in terms of a world existing independently of the subject. We do not normally feel that we see photons of light or "sense-data," but *objects.*[4] The distinction of "appearance" from the posited "reality" occurs on the most elemental levels of perception: by focusing our eyes on an object a few inches directly in front of the eyes and then noticing some object behind it, we can produce two images of the second object—but no one believes two physical *objects* are thereby created. On a more advanced level, since it takes light time to travel, we actually see a seven-to-eight-minute-old image of the sun (if present theories are correct), never the sun itself.

Our background knowledge is also brought into perceptions: we *see* that one person is farther away than another, not that there are two persons of radically different heights standing together—that is not the way things "really are." The world is never uncritically accepted as it appears. Seeing is not a simple act, but involves education; and thus people, not eyes or cameras, make observations. The same stimulus can give rise to different perceptions, as with such Gestalt figures as the Necker cube and the Köhler goblet/faces. The conceptual elements derived from previous habits, beliefs, and expectations are temporally prior to experience at each point. Cognition is not the discernment of regularities in an unadulterated stream of experience; rather, the stream itself is polluted at each succeeding point of its course by every prior cognition—and *consciousness* is reduced to only the faculty of responding to our own responses.[5] Our own conceptual posits thus regulate our perceptions.

Most scientists are not interested in the nature of observation but with what is observed, the reality invariant to differences in observers' particular experiences. Thus optics involves the wave-lengths of colors instead of the perceptions of normal and color-blind viewers—that is what is "really there" regardless of experiences. The much-valued *objectivity* of science rests on this desire for invariance. Such objectivity refers not to a world existing totally independently of all experiencing subjects—a belief shared by science and our common sense—but to our experienced world. Only those aspects of experienced reality which can be reproduced are scientifically significant. These are "natural," that is, reproducible or lawlike (whether in the everyday realm or in such bizarre realms as those of atomic phenomena or the interior of stars), rather than occurring without reason or as the result of a capricious supernatural agent. In effect, this objectivity consists of intersubjectivity and the specification of how the occurrence can be replicated.

Laws and Theories

The most elementary level of scientific systemization is the account of recurring phenomena in the form of *laws.* Under a realist interpretation laws do not concern sense-data nor are they merely economical or generalized descriptions of phenomena. Rather they delve into the workings behind appearances; they attempt to describe some structure of reality, not merely an accidental coincidence. As the understanding-component of science is given more weight, *theory* becomes central. Scientific theories are integrated systems of claims that are offered to explain the regularities noted by laws. They carry on the search for order amid the flux of experience at a deeper level of organizing structure. The process is to introduce theoretical entities, such as atoms and fields, that would account for laws and thus ultimately for appearances. Thus heat is explained by reference to athermal entities (molecules in motion); heat itself would not be understood if reference were made only to various hot objects. By incorporating entities that are unobservable in principle, greater systemization is possible: phenomena quite diverse on the surface are thereby seen as manifestations of the same underlying process, as with Newton's unifying the motion of tides, falling objects, and the planets as effects of one abstract construct (gravity). Fewer and fewer distinct assumptions then need to be made to explain appearances.

Theories thus are constructs—guesses at the deep workings and unities beneath the surface. As concepts reach more and more into the theoretical realm, connections to experiences and to causal understanding become more and more indirect. What is sought is a new framing of phenomena, new ways of looking at the world. Although no systemization is possible by simple induction from blind experience, imagination plays a greater role in formulating theories than in formulating laws. No set of rules automatically produces the new conceptual schemes. To guess at the workings beneath the surface necessarily involves creativity; concepts are formulated, manipulated, and reformulated into constructions that could account for phenomenal patterns. Concepts are rejected when they fail to make sense of experiences conveniently; new concepts are introduced to preserve simplicity in scientific systems. What is problematic in phenomena changes with the new theory. Formulating fruitful questions is difficult; judgments of how much weight is to be given each regularity that a phenomenon reveals occur at every step. But the theoretical framework is not an unrestricted creation of the human intellect; the constraints lie in ultimately having to account for specific states of affairs.

Testable claims (claims stating that such and such is the case, and giving directions for determining whether this is accurate) are the roots of the total theoretical network. Explanations must involve risk to be scientific: they must account for why one state of affairs and not another is the case. The least easily tested component of a theory, the one most difficult to subject to questioning, is the "ideal of natural order."[6] These ideals guide research by implicitly dictating what areas are worth investigating, what are the important questions, and what will be accepted as an answer. We *understand* only when a total theory conforms to this vision of the nature of reality. Examples of these well-entrenched principles: the Aristotelian notions that the natural place for heavy objects is the center of the universe, that circular motion is natural, and that what needs explaining is motion itself (why the arrow keeps moving once it leaves the bow). For Galileo, the situation changed: natural motion is movement in a straight line—this was "self-evident"—and what needs explaining is not motion but changes in motion. Today field theories have taken so strong a hold in physics that even though a classical Newtonian account of electromagnetic phenomena in terms of particles and action-at-a-distance was devised in the 1940s, no one has taken it up.[7] In biology the evolutionary point of view is accepted because no other theory in the field fits our vision of how reality actually is. The framework of evolution, as opposed to specific theories of its mechanics, is "indisputable" not because it is a simple, easily falsifiable fact that has been repeatedly verified but because of its position in our configuration of beliefs. What would scientists accept as evidence against a lawful continuity in the development of life-forms?

Scientific theories are advanced "tentatively," but once one adopts a conceptual scheme, spotting an assumption as problematic and arriving at a standpoint from which to question it are usually very difficult. It is not even easy to see that an assumption is an assumption until something new is thought up. Consider the great difficulty Kepler had in rejecting the idea of the "perfect" circular motion of Mars and in eventually coming up with a two-foci ellipse.

Explanation

Theories as a whole provide an explanation of the phenomenon under study by providing a conceptual framework in terms of which data will fit intelligibly alongside better-known data. The phenomenon is thereby "rationally explained," that is, placed within the framework of the beliefs deeply held at the time: what occurs ought to have occurred, if our beliefs are accurate.

Connecting explanations to specific observations has led some philosophers to tie scientific explanations closely to predictions.[8] Toulmin and N. R. Hanson have raised serious objections to this idea.[9] For instance, Darwin's theory has no predictive or retrodictive power of this kind but is still deemed scientific because it provides a way of ordering and understanding data via the concept of "natural selection." Conversely, Babylonian astronomy made very accurate predictions of eclipses but did nothing to further the current understanding of celestial phenomena. The process of finding an orientation or order to things has been more important in the history of science. Only if a theory is totally devoid of substantially accurate predictions does it become suspect; therefore one may question whether early Ionian astronomy, which ranks high in interpretative speculation, is indeed science.

Explanation cannot be a mere rewording of the problem but must supply insights which, to speak metaphorically, shrink the size of an unknown area of study by ordering what is known about it into some structure, thereby directing attention to smaller areas. This can guide further research; the outer planets and neutrinos are instances of discoveries resulting from theories' predictions. What is explained, it should be noted, are the observed phenomena, not the explanatory concepts. Those concepts invoked in explanations remain only partially understood or are deemed "self-evident." Newton, for example, conceded that he did not understand gravity: "I have not been able to discover the cause of those properties . . . it is enough that gravity does really exist, and act according to the laws which we have explained.[10]

Models and the Familiar

Rejecting explaining as equivalent to predicting also has implications for the connection of a theory's mathematics with models. There are many different senses of "model" in scientific usage: scale models, mathematical models, simplifying models, models as identical to theories, and so on. One important sense—a theoretical model—is of a metaphor developed between something familiar that is considered intelligible and limited aspects of a process or entity under investigation. Thus space is seen by Einstein as a three-dimensional counterpart of a two-dimensional surface of a sphere.

Are models merely heuristic devices whose "intuitive understanding" of a theory provides a systematic way to search for new laws (but are not part of the explanatory function of theories), or are they part of the explanation? Do mathematical equations alone exhaust the theoretical explanation? Newton rejected any fuller account of gravity than is provided by equations as "metaphysical poetry." Contemporary advocates of this view point to the consistent and successful mathematical equations of quantum theory, with its lack of any one visual model to handle all the experimental data. The opposite position is to follow what Pierre Duhem disdainfully called the "broad and weak" mind characteristic of the English:[11] understanding needs models or some sort of immediate conceivability, or we shall have reverted to the standards of Babylonian astronomy.

Whatever position is taken with regard to the role of models, explanation can be regarded as our attempt at understanding the unfamiliar in terms of the familiar[12]—conceptual understanding comes to rest only in ordinary experience, with that with which we are familiar. Even in the area of quantum theory, in the words of Heisenberg, the description in plain language will be a criterion of the degree of understanding that can be reached.[13] This criterion may reflect a permanent barrier to the amount of understanding we can have in this and other bizarre realms.

But science is also in tension with "common sense" (what is obvious or taken for granted by a culture or group at a given time). In every era science has challenged the common sense of the day: to Copernicus's opponents it was *obvious* that the earth did not move; their arguments against him had an almost *a priori* nature. Today examples of the unfamiliar proliferate: matter and motion affecting the geometry of the spatiotemporal continuum, mass increasing and length contracting with the increase of speed, and so forth. As Carl Hempel argues, it is the familiar that needs explaining by the introduction of quite unfamiliar types of entities.[14] However, it does seem necessary that some connection to something unproblematic or commonsensical at the time is required for an explanation to be accepted as satisfactory. (What is commonsensical does change as new scientific theories become more readily understood and accepted by a culture.) Common sense rules the setting up of experiments. And no matter how unexpected the results along the way, the end will have to be reconnected to the intelligible; otherwise the experiments would continue. The common sense of a period and its science stand and fall together.

Observation and Facts

It is essential to realize that theories are integrated units recasting laws and even the empirical phenomena in light of theoretical considerations. Understanding remains central. Thus, when a theory is rejected, the scientists go back to the drawing board of theory rather than just start new experiments. Without an initial idea, our observation would be aimless (although accidental discoveries do sometimes occur).

No absolute distinction between theory and observation is possible. Some terms are relatively theoretical and others relatively observational, but observations are not theory-free, and theories evolve in contact with observation. Observations are not simple experiences that are built into theories but involve seeing the *significance* of a phenomenon. Observations are relatively clear, quickly decidable, and agreed upon. Yet their integration into the structure of the theoretical network enables theoretical (unobservable) entities to be "seen." In normal research, scientists do not see droplets of water in a cloud-chamber; they see the tracks of subatomic particles. Only when a theory is challenged can scientists strip away the theory and view the data in a more raw state.

Theory also is fundamental in determining what counts as a *fact*. Even to call the earth a "planet" or to speak of other planets entails a theory-commitment—Ptolemy would not treat the earth as one of many planets. Determining how the world "really is" involves going beyond perceptions to the processes beneath them, and when this is done what a theory deems real become central. Observation reports may be dismissed out of hand because they conflict with accepted theories. Consider W. V. Quine's thoroughgoing pragmatism: the totality of our so-called knowledge and beliefs is a man-made fabric impinging on experience only along the edges; adjustments in part of the system affect the rest of the fabric, but any statement can be held true if we make drastic enough adjustments elsewhere within the integrated system.[15] Parmenides probably went the farthest in this regard by dismissing all sense-experience that conflicts with his reason. Today fairly well-attested reports of such paranormal phenomena as extrasensory perception are dismissed by many scientists for purely theoretical reasons. Hempel admits that this is rejecting the data rather than the "well-established" theory, even if no satisfactory explanation of the presumed error in observation is possible.[16] Or as one scientist put it: "This is the kind of thing I wouldn't believe in even if it were true."[17]

There are no "bare facts"—a fact is an event as we see it.[18] And how we construe events depends upon the theories we hold. As Karl Popper says, we formulate the questions we ask of nature and in the end it is we who give the answer.[19] Or from Hanson: facts are the objective organization of states of affairs within a scientific subject-matter which render true the theories we do hold; they are the boundary conditions of a theory: what is out there is accorded fact-hood because it anchors the least vulnerable parts of a system.[20] A corollary of this is that what we see as a *cause* is also determined by theory. Even to speak of "cause" and "effect" indicates a certain construal of experience. Causes are not given in sense-experience the way colors are, but depend on what needs explaining and what is deemed effective. "They [the causes] are not simple tangible links in the chain of sense experience, but rather details in an intricate pattern of concepts."[21] In a similar way "natural forces" and "rational solutions" are theory-dependent.

So also because of the role of theory, there are no "hard facts." The sciences are sometimes seen as giving us permanent, reliable, even certain, empirical knowledge—this body of facts only grows throughout whatever theoretical changes may occur. This, however, ignores the theoretical dimension of concepts. For example, our empirical knowledge concerning the sun has increased drastically in the modern era, but our understanding of its nature has also changed radically. What has remained "hard fact" throughout all scientific revolutions? In addition, what we construe something to be depends on how it fits into the puzzle of a particular theory. Different scientific disciplines and theories may construe the same phenomena as embodying different facts depending upon the limited purpose at hand.

Thus facts ultimately are constructs resulting from a scientific inquiry, not the beginning points. What was once theory may become fact in the development of our thought as the theoretical ideas become more widely accepted and inextricably interlocked with wider circles of our thought. The Copernican theory may be *fact* today because it has become part of the basis for much other research; it is no more observationally confirmable than it ever was, but it is so ingrained in contemporary science that it is hard to see it as other than indubitable. On the other hand, in the next century what we accept as "established fact" today may be rejected because it does not fit the theoretical networks of the day.

Theoretical Change

This discussion of "fact" raises the problems connected with the change of theory. The history of science is characterized by the interplay of continuity and discontinuity. The bias toward regularities is never an absolute screen and experiences in conflict with expectations do surface. If these anomalies are plentiful and significant enough, they lead to dissatisfaction with the theoretical framework. Changes usually come from younger scientists who are not so fixed in their ways of thinking. The continuity in theory-change comes from the fact that new theories result from reflection on the unexplained phenomena sticking out most prominently within the older theories; they determine the basic direction in which to look for new ideas. The discontinuity comes from the replacement of one theory with another: differences from ideals of nature down to facts result with the switch of conceptual bases.

The discontinuity of theories gets more attention in discussions today. It conflicts with the cumulative view of scientific growth, that is, that each scientist adds his or her brick to the wall of knowledge without affecting the other bricks or the overall structure. The analogy for the view popular today is that the wall of knowledge is periodically torn down—even the bricks are remolded—and a structurally new wall is assembled in its place. This emphasizes revolution in the history of science, but does not make room for the smaller conceptual changes that occur more frequently or for the continuity with the designs of the past. If, as is often done, the growth of science can be likened to evolution, then the Lamarckian version rather than the Mendelian theory of random mutation is more apt: this allows for traits acquired environmentally by one generation of theories to be inherited by the next.

Because scientific theories are integrated systems of concepts and propositions accounting for experiences, if a predicted phenomenon is not produced, it is not only one hypothesis that is in question, but the whole theoretical scaffolding used by the scientist.[22] The theory may undergo minor adjustments or its conceptual basis may be replaced *in toto*. Such changes in representation entail a change in all the concepts central to a theory.[23] What Copernicus meant by "the sun" is totally distinct from the Ptolemaic meaning (although the referent is the same in one important sense). So also the change in theoretical framework from classical mechanics to relativity theory is total: the Newtonian laws of motion are a very useful tool for a wide range of experiences, but they are *not* an approximation of relativistic physics since the frameworks are distinct—no amount of calculation with one set of concepts will produce the other. For Newton, mass does not change with changes in motion (and so force is linear), nor is there any reason to suppose that the speed of light is a barrier of any sort. The concepts of "space" and "time" are also altered. Similarly with the Newtonian and Einsteinian theories of gravitation: the former presumes the world to be composed of distinct bodies in space, which tug at each other thereby producing "action at a distance"; the latter presumes no force, no action at a distance, but a "field" (a "condition in space" potentially producing effects) that channels movements as ridges determine the movement of marbles. With the replacement of the Newtonian framework, the "attraction" of particles and "action at a distance" came to be seen as useless concepts, as Leibniz had felt them to be. In sum, for these two theories the make-up of the world, its basic features, is different. Since the facts are determined by theory, holders of different theories in one sense live in different worlds, the world being the sum total of facts. And the phenomena will be sliced up differently. But different theorists also live in the same world: they do not speak past each other in accounting for the same stimulus and its source.[24] To use an analogy to a Gestalt figure, one cannot erase the "goblet" without also erasing the "faces."

Grounds for Theory-Acceptance

When new music is played, there often are disgruntled listeners—riots have even occurred in normally sedate concert halls. The comparable problem here is that holders of old theories often have trouble even understanding new ones. But if this hurdle can be overcome, what counts as grounds for accepting one theory over another? Sense-experience is of course central in any appraisal of a theory. But its role is not one of straightforward confirmation or disconfirmation: fulfillment of precise predictions cannot prove one theory rather than another accounting for the same data. Positive results only "affirm the consequent." Repeated confirmation does not improve the situation logically since induction plays at best a minimal role in the evaluation of a theory: a thousand confirmations of a theory are statistically insignificant compared to the infinite number of possible tests. Conflicting future results are always possible as are other theories that can account for experimental "confirmations." Imre Lakatos, following Popper, points out that the best corroborated scientific theory of all time was Newtonian mechanics and gravitional theory: the conceptual transformation to relativity shows that scientific truth is not provable or even *probable* on predictive grounds.[25] Thus it becomes difficult to attach probability in a mathematical sense to the possible truth of the theoretical dimension. So also faulty theories can lead to new discoveries, as with Newtonian physics and the planets. Another historical example illustrates another problem: the "overwhelming" evidence for the interpretation of electromagnetic theory with reference to ether did not convince scientists of the "physical reality" of the elements that rendered the theory intelligible.[26] Thus verifying evidence cannot guarantee a theory's truth.[27]

Even if it is not sufficient, accurate prediction has been advanced as a necessary condition for the acceptance of a theory. Popper's famous falsificationist criterion for delimiting what is scientifically interesting from what is not is such an idea: to be scientific, a theory must specify beforehand what observable situations, which if actually observed, would mean that the theory is refuted. But the problems are just as great here. (1) There is no simple falsification of a theory. Theories are not generalizations that can be overthrown by finding one negative instance; nor can probability-claims other than certainty be refuted by negative results. If this were the case, all scientific theories would have to be rejected because all have recalcitrant anomalies surrounding them. (2) As mentioned earlier, theory can overrule experience. The Copernican theory blatantly conflicted with the obvious experiential situation, that is, there was no indication that the earth moved and there was the problem of the lack of parallax; and its predictions were no better than those of its rival. This caused Galileo to speak of Copernicus's triumph of reason over the obvious. If a theory looks promising, not only initially negative results but also enduring anomalies are dismissed until the indefinite future. (3) Some ideas that have appeared to have been banished forever (e.g., atomism and heliocentrism) have returned many centuries later. And (4) all theories confront experience, not as a series of isolated hypotheses, but as a unit. When negative results occur, choices remain as to which part of the theoretical network needs to be modified or abandoned. The broader a component is in scope, the more important it is and thus the more resistant to falsification. Thus the Michelson-Morley results were never taken as evidence that the earth does not move. Some people would even argue that there are no "crucial experiments"; since any experiment involves more than one assumption, a single assumption cannot be isolated from all assumptions to be uniquely tested.[28] Sometimes a series of theories—each requiring justification—separates our experience from the theory in question. Or anomalous experiences may be handled by secondary elaboration that is branded *ad hoc* from the point of view of other theories. Even Popper himself concedes that no conclusive disproof of a theory can ever be produced. A theory is too complex an entity to be established or refuted by observation alone.

Scientific theories do ultimately have to account for experiences and therefore are constrained by them. Yet from the above discussion it appears that Toulmin is correct in questioning whether sense-experience alone ever serves as *evidence* for or against any scientific position.[30] Theories are underdetermined by all observations. Experiences can motivate a decision for a basic statement, but cannot justify it.[31] The empiricist ideal that the ultimate criterion of truth in science is theory-free sense-experience is in error. Support for a theory, and thus ultimately the facts it considers as making up the world, must come from conceptual as well as sensory sources.

In effect, there is a mixing of the conceptual and experiential in the acceptance of theories. Intelligibility once again is central: experiences are signs or clues to the workings of nature but are insufficient in themselves to provide understanding. On the most sensory level (e.g., measuring the sides of the Necker cube, as opposed to deciding if the side is in the front or the back), observation determines the results more straightforwardly; falsification at this level is at its strongest. But when it comes to determining what the *facts* are (i.e., placing the pieces in an intelligible pattern), considerations of fundamental structure—the ideals of natural order—which are not directly decided by experience come into play.

Criteria proposed for theory-acceptance other than empirical accuracy and predictive success include simplicity, internal consistency and systematic organization, coherence with other accepted theories, scope (how many different types of phenomena are covered), fruitfulness, familiarity, mathematical elegance, and the intuitive plausibility of the ideals of natural order. Objections can be raised to each one. Take simplicity. It is assumed that the simpler the solution the more likely it is true because reality is constructed simply. But does this economy merely reflect a human need or taste? Simplicity certainly does not hold absolutely; in metaphysics, solipsism after all is the ontologically simplest position, but few people defend it. In science, more complex theories may overrule simpler ones on other grounds. The simplest explanation of paranormal phenomena—that there are real paranormal powers—is not readily accepted by many scientists today. Nor are any of the other criteria absolute: the fruitfulness of Newtonian mechanics in some areas (finding planets, putting people on the moon) does not overshadow problems with it. Coherence with other theories has a conservative effect: revolutions in theory need to overcome such resistance. Mathematical elegance and the beauty of the general framework combined in the case of Einsteinian relativity for scientists to dismiss damaging experiments as being the results of unknown sources of error. The general vision of reality provided by a theory makes it seem "self-evident" or "reasonable to believe" that, say, stars are far away or galaxies are moving apart. But such visions of reality are themselves replaceable. None of the desiderata is as central as empirical accuracy, but even this is not sufficient and can be overruled in defining "facts." And no algorithm exists for mixing the criteria in devising a theory's justification.

Anomalous experiences count negatively against especially broad theories only when another theory is advanced that covers the same phenomena, accounts for at least some of the anomalies of the older theory, and looks promising.[32] Only then does the weight of complications cause a theory to collapse. The new theory usually makes predictions different from those of the previous theory. New problems may arise, but anomalies are not reciprocal: the theory that wins acceptance accounts for some outstanding anomalies of the loser, but not *vice versa.*

Since sense-experience alone cannot select scientific concepts and no rules churning out justifications exist, judgments must be made. Judgments made at the frontiers of research and at the metaphysical foundations are of most value to the future direction of science. Such judgments are social and public in nature, since a community of specialists decides, but they still involve weighing variables in the construction of a conceptual apparatus. Is the choice between theories *rational?* A remark by Max Planck would indicate that the advance of science is nonrational: "a new scientific truth does not triumph by convincing its opponents and making them see the light, but rather because its opponents eventually die, and a new generation grows up that is familiar with it."[33] Thomas Kuhn and Paul Feyerabend also raise doubts. For Kuhn, after a revolution scientists respond to a different world. Rival groups do not share all assumptions, and their theories are incommensurable in the sense of lacking any common ground for the adjudication of disputes. Each theory embodies its own standard of rationality. And since the theories are built into perception, the different groups see different worlds.

There may be a common ground, though. For example, despite the fact that all perceptions are "theory-laden," they are not necessarily laden with the theories in dispute—agreement on less ramified concepts (e.g., "celestial body" instead of "star" or "planet") is then possible.[34] Thus it makes sense to say that different theories can cover the same phenomena, and to speak of a genuine choice between theories. Kuhn, in later clarifications, also stresses that there is the possibility of partial communication between holders of conceptually incommensurable theories: one can understand without believing. So too strong arguments can be made at first for both the old and the new theory. But Kuhn does concede that decisive arguments are developed later: good reasons in terms of predictions, simplicity, compatibility with other theories, and so forth, are advanced even if judgments and persuasion are involved.

Feyerabend, the self-proclaimed Dadaist of epistemology, goes farther in claiming that such choices are irrational.[35] Sometimes he means only that no amount of marshaling of evidence and reasons can logically compel scientists to change their minds. This is true: one, in his words, cannot rationally criticize a scientist who sticks tenaciously to a degenerating research program, because a scientific butterfly may emerge when the caterpillar has reached its lowest stage of degeneration. Scientific progress is not always achieved by the open-minded: sometimes one should forget one's doubts and press on; one cannot be free of theoretical prejudices and can only hope one has the right ones.[36]

All the reasons for accepting a theory will not be covered by sensory and logical considerations alone—aesthetic judgments, judgments of taste, metaphysical prejudices, and religious desires play their role in theory-change.[37] Feyerabend, however, makes this situation sound less sane than it is when he speaks of "brainwashing" and "propaganda." Ultimately, *no* decision is totally rational.[38] Choices and arguments, not logical deductions, remain. But even if reasons acceptable to all disputants cannot be advanced for accepting a theoretical framework, still discussions are possible concerning how much weight to give each criterion for theory-acceptance and problems with other theories. The reasoning a scientist employs to do this is closer to that of a court of law's judgment rather than to a conclusive logical deduction.

For this reasoning process, factors from the culture of the time and place enter into what is scientifically acceptable. In fact, what is considered knowledge in a culture enters the entire process of science: education and culture condition stimulus in the process of perception, select what is important, and determine what is in need of explanation. Even intuitions are not *sui generis* but arise only when the mind is resting after intensive study that involves the wider range of one's total background. Models come from what is familiar and commonsensical in a society. Simplicity, scope, and so on depend in some degree on cultural standards. Explanations are relative insofar as they are always making sense of a situation for a particular group in a particular historical-cultural context. Thus it is difficult, if at all possible, completely to separate the contexts of "discovery" and "justification": the cultural factors operating in the creative devising of new conceptual frames operate also in the accepting of them. The "rational" justification of a claim involves connecting it to the general view of the world from which it arose. Sociopsychological causes may be responsible for why a new theory gets an initial hearing. For example, Darwin's evolutionary theory fitted well with the capitalism and sense of progress in Victorian England. In the longer run, more reasoned considerations figure in the acceptance of a theory, but cultural factors remain part of the picture.

So far *truth* has not been mentioned in this discussion of theory-acceptance. Kuhn believes that science is the surest example of sound knowledge we have and that science is progressing (although he does not specify in what manner). But he says we must relinquish the idea that changes in theories carry scientists closer and closer to *"the truth."*[39] Theory-change occurs through natural selection, with no progress toward any goal. Each is a momentary adaptation following another, with no preordained pattern leading to a final answer. We make the puzzles and we solve them: there is no "real" solution.

This may be correct. Certainly unanimous consensus among scientists concerning one theory is no guarantee of truth. Ptolemy and Newton after all had unquestioned support for centuries. But it is not necessarily the case that there is no final account of the world open to us. Science has been progressive in opening up a wider range of phenomena to us. And there has been progress of theory in the different fields: a wider range of phenomena has been accounted for more simply and the empirical "fit" has been better (fewer phenomena are seen to be exceptional, or only more esoteric phenomena appear anomalous). Thus there appears to be a convergence on "truth" in the sense of a simple account of the workings of the world. A final account may never be reached, since there may always be room for conceptual refinements, but progress still might occur.

On the other hand, there is always the prospect of a major conceptual revolution. Certainly the Copernican, quantum, and relativistic revolutions have shown that what is taken for granted may very well be radically changed, and thus that all theories must always be held as tentative. We may be reaching the "perfect" scientific theory in each domain of study—the simplest by all standards, the most intuitively obvious, and so on. But unless this kills off all research, the prospect of replacing the foundations of all present theories remains. What was previously thought to be completely understood may be given an entirely new basis. Or there remain choices at the frontier and foundation of science. Hesse notes that the succession of atomic theories exhibits no convergence in the description of fundamental particles—there is an oscillation between continuity and discontinuity, field conceptions and particle conceptions.[40] Consensus may occur or Feyerabend may be correct in advocating a pluralistic science, that is, various theories covering one domain, each held tenaciously and each giving a different insight into the nature of the world. No one systemization of sensory experience may prove the best.

Science and Metaphysics

Most scientists would be reluctant to forgo the notion that science leads to truth. This conviction is a metaphysical presupposition of science. Science is not tied to any particular metaphysical system setting up a systematic and coherent conceptual scheme covering all types of experiences. Science is, for example, neutral as to whether phenomena are crystallizations of an ultimate underlying consciousness—its commitment to the structures as real is compatible with this metaphysical position concerning the realm of structures or its opposite. Nor is science a form of *naturalism*. But in making claims about the world science is dealing with aspects of reality. Theories may divide up the world along certain lines in order to direct attention to certain patterns or regularities without arguing in addition that the world consists solely of the elements detached for the purpose of study. When Galileo differentiated "primary" from "secondary" qualities along the lines of whether an aspect of the world was quantifiable or not, he was making a scientifically legitimate move: science is interested in those intersubjective features which do not vary from person to person. There are measurable and nonmeasurable features of the world, but to dismiss what is not measurable as the realm of opinion or illusion is to make a metaphysical move that science alone does not warrant. Noting that what is, say, causally effective is real is one thing; saying that it is the essence of the phenomenon or the sum of reality is another. A method or research program need not be made into a metaphysic in this sense.

There are senses, though, in which it is useful to speak of metaphysics in science. First, the ideals of natural order incorporated within the theoretical framework—the "first principles" of a "natural philosophy"—border on speculative metaphysics because they define the nature of the phenomenon in determining what will be accepted as a fact, a reason, and an explanation. They set up frameworks of intelligibility far enough from simple falsification that it is difficult to demarcate absolutely science and such metaphysics. Second, there is also a metaphysics of science, that is, broader ideals of natural order that are presumed by all scientific theories and thus underlie science as a human enterprise. Science as a whole presumes a faith, as it were, about the nature of reality. These articles of faith are not testable hypotheses nor premises, even implicitly, within specific theories. Rather they are the tacit components of the scientific framework within which all scientific research takes place. To give up this framework is not to disprove one specific theory but to give up doing science. These metaphysical assumptions about the nature of the world are also general rules insofar as they determine what questions can legitimately be asked and how the data will be interpreted. None of these articles are necessarily unique to science, but this does not change the fact that to give them up is to give up science.

(1) One article is realism in the sense of the denial of any solipsism: any discussion in science presupposes without question that there is a world existing "externally" to our minds and "behind" the appearances that are partially dependent upon ourselves. Scientists in quantum physics, where the measuring apparatus affects what appears, and in relativity, where some of an object's properties depend upon the frame of measurement, have no more reason to question the independent existence of a world than do other scientists.

(2) This external world is divisible into parts (entities and processes) for analysis without a distortion of its nature. Scientists usually presume distinct material objects, but science does not presuppose unchanging metaphysical substances. The focus instead is upon structures.

(3) The world is rational, that is, it is intelligible and comprehensible to us in its physical structure. That sense-experience can give us accurate information is part of this, as is the assumption that the rationality embodied in the process of justification of scientific theories reveals something of the structure of reality even if it does not exhaust it. We may not be able scientifically to understand aspects of reality, but this article of faith indicates that we would treat this situation in the physical and biological aspects of reality as only a temporary block in our ability, not as an inherently incomprehensible and yet "natural" process.

(4) As previously mentioned, truth is an article of faith. Science may not

be the only source of truth about the world, as some metaphysical positions allege. But reliable scientific constructs, although simpler than complex reality, are felt to convey some accurate information about the world.

(5) Nature is orderly and uniform, that is, under the same set of conditions, the same phenomena will occur. If reality were not uniform, if each event were absolutely unique (involving no repeatable features), no science would be possible. Nor could science proceed if the world is seen as a cluster of chance coincidences. Order is presumed in explanations and understanding. Is this belief in order the result of investigation rather than being presupposed? If this were so, something could occur that would convince us otherwise. What could occur that would convince us that the order we find is only our own habits and prejudices projected upon the world, that is, that we see only what confirms our order and the rest is chaos? If nothing in principle could count against this belief, it is not an empirical conclusion but a regulative principle.

This presupposition does not preclude the possibility that regularities may change throughout the history of the universe or that they may be very different in other parts of the universe (e.g., in the interior of stars). Because some conditionals are set up without temporal or spatial referents, this does not entail that the future and the distant past must be like the immediate past or that the whole universe must be like the parts we experience. We may find otherwise through investigation. Our laws may have limited scope and the conditions of applicability of each would need to be specified. Nevertheless, we assume that when the same conditions arise the same phenomena appear.

(6) The workings of nature are law-governed and can be explained by reference exclusively to natural, law-governed events. Nothing supernatural (capricious, unlawlike) needs to be appealed to. Only natural factors are required for understanding, and no event is exempt from this understanding.

Each of these articles may be mistaken. Faith does involve risk. But to give them up would be to end science. Challenges such as the apparent randomness of beta-decay does not end research into the subject. Since science has such a strong hold on our beliefs about the world, it is hard to imagine even our ability to change these articles. We may not be able to take a standpoint apart from science from which to criticize.

The Accuracy of Scientific Claims

There are a number of epistemological issues concerning the soundness of scientific claims that still have to be raised. These involve the fact that scientific knowledge is an interaction between people and a reality independent of them. How much do *we* contribute to scientific knowledge and how much is contributed by the *reality* observed?

The first issue is conventionalism, the position that our image of the world is not uniquely determined by experiences and that we are free to choose whatever conceptual stylization is convenient—only when a conceptual map is adopted do experiences play a role in determining the truth or falsity of a claim. It is true that we construct schemes along the lines of what is convenient to us. As more and more experiences occur, new schemes may become necessary, previous ones no longer being fruitful. And experience alone never selects the "proper" scheme. We select from alternative conceptual systems the one that for the time being makes the most sense out of those regularities we note. These are conventions imposed on our science, not on nature.[41] Our theories and images of the world are like fish nets, to use Arthur Eddington's analogy, sweeping through the sea of experience and catching only part of what is there. We must be aware that the method employed determines the data in the sense of selecting for study only part of what could be seen. It does not determine what is caught in the sense that the phenomena within the area of study are there or not. We choose the nets, with their inherent limitations, but use them as tools to find something not of our own making. Conventionalists are correct in emphasizing this role of the scientific theory, but (as will be discussed shortly) whether the fact that some nets are more convenient than others in itself also reveals something of the nature of reality is something conventionalists must deny.

In one sense, however, how we arrange the regularities does determine what facts we see as making up the world. This involves the choices at the foundations of the theoretical networks. For instance: when Einstein developed relativity, he saw that Euclidean geometry and the classical laws of mechanics and optics could not both be maintained. He opted to give preference to the laws of mechanics and optics, and to vary the geometry. He could have done the opposite—Henri Poincaré thought the classical geometry held too central a role in our thought to be abandoned. But whether light is bent by gravity in Newtonian space or follows "straight" Riemannian lines is in the same class of questions as whether the earth moves, namely, questions of the structure of reality. Predictions and observations may be unaffected; nevertheless this does not constitute equivalence of theories.

Another issue is the ontological status of theoretical entities.[42] All positions on this matter presuppose the realism of a world existing independently of the experiencing subjects. The question is whether theoretically unobservable entities make up part of the real world or are merely artificial constructions introduced for organizing intelligibly the flux of experience.

Eddington represents the idealist response to all epistemological questions. Mind fits nature into a framework of law largely of its own choosing, and discovery of this system is merely the mind discovering what the mind has put into nature.[43] Nature is neither a cosmos nor chaos: we superimpose laws and meaningfulness. There is no order independent of our minds and we confuse our own constructs, including theoretical entities, for the structure of reality. This position, however, is not supported by many, since it goes against science's metaphysics.

Another position, the instrumentalist, presents theoretical entities as human inventions ("fictions") facilitating inferences from one group of phenomena to another. Laws then are not descriptions or summaries but *rules.* They are, in Toulmin's phrase, "inference tickets" that hold or fail, rather than being true or false; only statements of their range of usefulness are true or false. Scientific objects are not real but only conceptual links between what is real (the phenomena).

Last, the realists: theories are true or false in matters concerning reality; they would not be successful rules of inference or fruitful sources of new interpretations unless they touched something existing independent of our experiences. For realists, the claim that the earth moves is true, and not just the basis for a convenient conceptual system. And theoretical entities are actual entities, not merely useful devices within a conceptual system, about which science is steadily learning more. This is not a commitment to naive realism (science describes the world-in-itself rather than as experienced): science is the result of the interaction of the observer and the rest of the world; to this process the world contributes structures existing independently of us that are touched upon in theory.

The strongest argument for realism is that most practicing scientists do currently speak of correctness and truth, not just convenience. Doubts can be raised, though, concerning the accuracy of theoretical constructions. The first arises from the fact that science is a *human* enterprise. Theoretical entities are our posited constructs, which are only guesses at the processes operating beneath the surface phenomena. Theories do seem to strike some features of reality—success in predicting new patterns and regularities is most easily interpreted as indicating that. But the role of our imagination almost certainly introduces error: our understanding is not identical with what is there. Copernicus mixed the idea of the earth's moving with the errors that the sun is immovable and the center of the universe. Theories have been advanced and discredited regularly. Theoretical entities such as "ether" have been introduced, been accepted, become too complicated, and died. The intellect does not necessarily distort what it touches, but it sees incompletely. Thus, strictly speaking, all our theories are false.[44] And as our constructs move deeper into theory, they become farther removed from experience. Anomalies abound around any theory; still, this does not rule out the possibility that science might tentatively disclose features of reality.

Levels of Organization

One final attack on the possible accuracy of scientific claims is the thesis that different types of scientific theories contradict each other—for example, if physics is correct, biology is wrong. The systems approach to studying nature provides a response: nature consists of a hierarchy of levels of organization each of which has properties of its own. From the smallest realm of interactions, levels lead from reality without chemical properties up through viruses to the animate realm and through aggregates in general to cosmological levels. Each "whole," whether a nucleus, a cell, or a galaxy, has properties of its own that make it more than the sum of its "parts." Parts in context have properties that do not emerge when the parts are in isolation. An atom is not just a pile of subatomic entities nor a cell a pile of chemicals: each has an internal structure that makes it a whole, a reality whose effects cannot be accurately analyzed into its component parts. Analyzing hydrogen and oxygen individually will not reveal the wetness of water, as Niels Bohr said. A person cannot be made by assembling a pile of the appropriate chemicals and stirring it up, nor can a person be completely analyzed by any analysis of these chemicals. How much a human body is worth financially depends upon the level of analysis involved: on the simplest molecular level, it is not worth much; but on more complicated levels of organization (cells and organs), the parts become quite valuable. Such structure in addition to the "stuff" structured is real too, insofar as it must be taken into account to understand a phenomenon. The mechanistic approach to nature (cutting reality into bits and analyzing the relation of one bit to the next) works with varying degrees of success depending upon the type of organization within each level of reality; it in no way accounts for the relation between the different levels.

Does reductionism run counter to this? No. Reductionism in science is the attempt to derive the laws of a discipline studying one level of organization from the theory accounting for another, more "fundamental" level. The *reality* of each level is not questioned, but only the explanation.[45] Concepts and laws of each domain are irreducible in the sense that the laws governing cell reproduction will not be found by analysis of the constitution of an atom's nucleus. A reduction occurs when conditions of the parts can be specified that are necessary and sufficient for the occurrence of a property of the whole; the property of the whole is no less *real* merely because its occurrence can be predicted. (Whether there are any successful examples today of reductionism in science is doubtful.) Each level still has its own properties and requires its own analysis; descriptions may be complete on each level.[46] Nothing in the scientific approach attempts to explain why reality is organized into a hierarchy of emergent levels.

Behind the desire for such reductions lies the metaphysical position that only matter with the most elementary structure, or totally free of all structure, is ultimately real. Science *per se* is not tied to such a philosophical position and cannot be used to justify it in any simple deductive manner. One problem with this position is that the phenomena in every level so far uncovered are no simpler than those on higher levels—no simple level of the ultimate constituents of the universe describable in a few simple ideas has been forthcoming.[47] This raises the prospect that there is no such level and that the metaphysical desire will go unfulfilled. That is, science deals with patterns, not the "stuff" patterned; whenever the focus of attention is sharpened, "matter" dissolves into complexities—matter, in effect, is just patterns out of focus,[48] and this will remain the case however "deep" we go.

A more important difficulty with this metaphysical position is that the reality of the structure cannot be denied. Each "lower" level may define the conditions for the emergence of the next, but this does not mean that the lowest level is reality "as it really is" or the sum of reality. Quarks (or whatever) may contain the conditions for the emergence of all other properties in the universe, but to study those properties manifested only through the interaction of quarks with each other will not reveal these properties. Each level is real; the universe is organized into subsystems up to the universe itself as a whole, and no one level can be selected as the sum. Since scientifically all levels have an equal claim to reality, choosing one as the fundamental master structure of all reality will depend upon our wider conceptual attempts to develop a simple coherent system. That is, the everyday level is as much a "cause" of subatomic particles as *vice versa* unless we have a metaphysical viewpoint that dictates otherwise. This is not to endorse vitalism (the position that living organisms involve a separate entity or force) or a proliferation of incompatible scientific entities: each whole is not materially more than the sum of the parts. But the structure, even at the lowest levels, cannot be denied.

Each theory in each scientific domain is a net for a different purpose, each catching some of the world. Each accounts for only certain aspects of the world, depending upon whether the scientist's interests are in biology, geology, a branch of physics, a hybrid discipline, or whatever. Consider the various causes of a person's death in a murder case: from the legal point of view, the murderer is the cause of death; from one point of view in physics, it is the gun's ability to propel the bullet; from the physiological point of view, the bullet's damage to the body caused the death; and so on. Each event is a complex of many such causes, and each science treats certain of them as relevant to its domain. There is no one final scientific picture of the world—only different pictures for different purposes.[49] No one description or explanation exhausts the way the world "really is." Each gives some insight into a limited aspect of phenomena. Each of the various physical sciences covers everything that is real, but only aspects of it.[50] In discussing quarks we are speaking of all of physical reality; the same is true with relativity. But in neither instance are we discussing all levels and structures of reality: physics does not have the concepts to deal with biological or other levels. Each "fact" is an abstraction serving a special purpose.

Nor is there a conflict between the everyday realm and the scientific realm as a whole. Science does not substitute another world for what is experienced but rather accounts for phenomena on particular levels. Scientific constructs are advanced to explain regularities in sensory experience and need to differ from what is to be explained to accomplish this—ultimately to explain color, we need to invoke something without color. What is true in one context needs to be false in the other. Still, the world is one, but only studied from different points of view; there is no causal relationships between processes of different realms. Some properties (e.g., solidity) occur only on the level of everyday aggregations: their parts do not have the properties individually. The concept "table" may be just a convenient construct simpler than the literal truth only if all its properties can be shown to be properties actually of its parts; but if the parts have properties *in situ,* a table is a reality too. Everyday entities may not have the properties of what is labeled the most fundamental level of reality, but at no point is it a scientific assumption that the experienced world is not real—or that the everyday realm is the only realm of reality. As far as the scientific enterprise is concerned, the world "has" or "is" the properties revealed at each level. Everyday entities and theoretical entities do not differ in this regard in any way.

2
Mysticism

(1) Mysticism and Religion

Religious Ways of Life

When we turn from science to mysticism, differences strike us first. The context for both of the Indian traditions under consideration here is one of religious ways of life. A "way of life" is a set of values, action-guides, and belief-commitments roughly integrated into a unit. Religious ways of life are such designs for living oriented toward an "ultimate concern"[1] revolving around a conception of fundamental reality, the ultimate powers controlling our life and regulating phenomena, or the ideal state of a person. Religious concern begins with the problem of meaning and meaninglessness—that we are out of step with reality. Religious systems provide the soteriological means and goal correcting this problem. The dimension of ultimacy gives religion a significance for the whole of life; it is not relegated to one smaller area of psychological or social concern or to special practices such as worship or meditation alone. Religious ways of life provide the broadest contexts of meaning for their practitioners and answer questions dealing with the ultimate significance of phenomena and experiences. Within their framework, life is meaningful: believers have an orientation and view of reality for coping with the hardships they encounter.

Ways of life contain beliefs about the world that make the other components (religious goals, emotions, values, actions, etc.) appear reasonable or even possible.[2] Belief-claims are not the sum of a religious way of life, nor its most basic component; but they embody the conviction that the practices and values one holds are grounded in the structure of reality.

Religion differs from science in the dimensions of significance (personal, impersonal, and "ultimate" significance) attached to all experiences and in the types of experiences (mystical, visionary, historically unique, etc.) deemed central. But it is assumed here that, like science, religion is a human attempt to understand reality. Religious systems are never advanced tentatively as a set of intellectual hypotheses explaining a problem as are scientific theories. And since a specific religious scheme covers a wider range of experiences and dimensions of significance, it is harder to uproot—it tends to overrule *prima facie* problems rather than *vice versa.* Its belief-component explains an experience's most comprehensive significance, not merely its various physical or biological types of significance. Ultimately, however, understanding and explaining in both science and religion consist in making phenomena intelligible by fitting them into an acceptable coherent conceptual scheme.

A distinction should be made between religions and religious experiences. Within different religious ways of life, all moral, scientific, or other experiences are related to that religion's understanding of the ultimate significance of our lives. There are also distinctly religious experiences, experiences whose religious significance is so overwhelming that other possible levels of significance are dwarfed in comparison for the believer. Any experience that transforms a person's life could be called "religious," but following the characterization given here, only experiences related especially closely to an ultimate concern (whether tranformative or not) will be deemed religious.

Mysticism and Mystical Experiences

The locution "mystical experience" has been used to cover a wide range of phenomena from parapsychological experiences to mythology. Here it will refer to two types of experience: the range of sensory and introvertive experiences in which the grip of concepts upon perception is weakened to one degree or another ("nature-mystical experiences"), and a "deeper" type of experience occurring when the sensory-conceptual apparatus of the mind is in total abeyance while one yet remains aware ("depth-mystical experiences").[3] "The mystical" is the reality allegedly involved in the latter experience. "Mysticism" is any way of life oriented around mystical experiences; and when an ultimate concern is connected to the mystical experience, the mystical way of life is also religious. Mysticism is not the essence or only form of religiosity; forms of religion exist that attempt to reject mystical experiences entirely (e.g., Karl Barth's form of Protestantism). Nor are all mystics part of "organized" religious traditions. But mysticism is usually religious, since the mystical experiences are most often connected to the "ultimate concern" of the mystic.

If there is one feature unanimously stressed by mystics, it the *experiential* nature of mysticism, that "enlightenment" is not merely a new interpretation of experience and phenomena arrived at through normal thought-processes.[4] Nor is enlightenment an intuitive "jump" ending a line of reasoning; rather there is a sense of presence, power, and reality. So also the two types of mystical experiences are substantially different experiences, not merely varying interpretations of similar experiences. A unique problem in studying mysticism is that probably every claim asserted by a mystic has been advocated by nonmystics for reasons other than those connected with a mystical experience. For instance, Hume speaks of the unreality of a permanent self; and Parmenides argues "all is one" for nonmystical reasons. The total systems rather than isolated fragments, however, usually make it clear if another mode of experience is being given central importance. Thus it is not usually difficult to tell if the concept "being" is merely the philosophical abstraction obtained by thought of what is common to all entities, or is the subject of a nonordinary experience. As with scientific theories, mystical utterances form an integrated system, not a series of isolated claims to be judged individually, and the importance of the mystical comes through the final product.

To see why mystical experiences are taken to be insights into reality, what mystics contrast with these experiences needs to be discussed. The deeper variety of mystical experience is totally distinct from normal sensing or mental activity where "normal" denotes broadly any activity involving the duality of a subject-object framework, whether this activity is everyday observation or the most highly refined scientific theorizing. Analysis and division are central in our need to survive. We concentrate upon fairly stable recurring patterns by means of which we render the actual flux of experienced phenomena intelligible and manageable. We select, usually unconsciously, those aspects of the flux which we deem most significant and enshrine them as real. The most central such construct is the "I," the sense of a conscious entity set off from the world "out there." Much of our action is then the attempt to manipulate the rest of reality for the benefit of this image of a center of mental functions.

Perception itself is a good example of the process. It is not a passive process—an impressing of an external world upon our brain producing a picture of reality. And a perceived "thing" is not a duplication of the stimulus that strikes the retina, but is a product evolving through layers of neurological, cultural, and personal filters. As with the rest of the social world, habituation is the principal agent filtering perception. Our habitual responses in the past set up a frame of expectation for present experience: by comparing incoming stimuli against what is remembered, only what is deemed important is passed on for further processing.[5] If the habituation is strong enough, only the categories formed by past experiences are experienced: stimuli activate an internalized category, and from that point on, we deal only with the category. As cited in the last chapter, consciousness in this way becomes only the reaction of our mind to our own reactions.

The role of language is essential to our ordinary awareness. Language is necessary to anything recognizable as human thinking, that is, the manipulation of concepts. More generally in fact, language "is both the foundation and the instrumentality of the social construction of reality."[6] This process requires distance between us and what is perceived, and a separation of elements in the flux of phenomena (i.e., giving them a status distinct from their surroundings). Language becomes the medium that stabilizes reality for our use by denoting those aspects of the flux which are essential to our advantage. *Naming* is the foundation of the perceptual/cognitive order, since only what has a name, which is encoded, has the status of being real. This concretization of the phenomenal flux involves categorization whereby phenomena become mere instances of broader categories. No vocabulary denotes each instance singly. All sensory stimuli are filtered through the habitual lines for slicing up reality enshrined in the conceptual organization and terminology of each particular language.

Our attention is normally tied exclusively to those conceptualizations that we construe as being the real world. According to the two Indian traditions that will be dealt with here, the suffering we experience is the result of operating within this manufactured world without knowing its status: we desire and cling to our constructions, which were initially introduced only for convenience. We create a world of multiple objects when in reality there are no distinct independently existing objects. Because the realm of "becoming" is actually complex and constantly changing rather than consisting of simple, easily manipulable pieces, the result is inevitable dissatisfaction. Our energy is devoted to purposive action consisting in the reaction to images, and this will prove frustrating when reality does not match our pictures. Mystical freedom can be understood at least partially as release from our conceptual cocoon.

To effect this freedom, the sense of a multiplicity of real entities must be overcome. Each mystical way of life attempts to achieve this by prescribing an orderly movement away from self-centered awareness. Alternative action-guides are established initially to reorient our behavior along more selfless lines. Mystical ways of life also contain special practices to "unlearn" our normal way of knowing. These are receptive and concentrative techniques leading to the deautomatization of our habitual patterns, and thereby to the avoidance of ossifying our social worlds into illusions. Basic awareness is freed from the dominance of our habitual anticipations and categorizations. This occurs either as an aftereffect of the depth-mystical experience or as constituting the nature-mystical experience itself. More "raw sensory data" come through when the processing and abstracting mechanisms are weakened.[7]

What result are nature-mystical experiences, that is, an awareness not dividing the world up into distinct, self-contained units. The very idea of a "world" separate from a "self" and filled with independent "entities" appears from such experiences to be an *illusion* generated by our misreading of the nature of sensory data. What we conceptually separate as entities are eddies in a total field of integrated "becoming." The extreme instance on a continuum of possible nature-mystical experiences, the one farthest from ordinary sensory experience, results from a purely receptive mindfulness—

receiving sensory stimuli without any conceptual structuring. In other instances there may be a sense of union, of being one with the whole, but there is also a sense of different nexuses within the flow.

With the depth-mystical experience, the emphasis shifts from the realm of change to the still center of being—an immutable reality "behind" the changing flux. In nature-mysticism, if anything is constant it is the total whole of phenomena. In depth-mysticism, there is a oneness that is not the totality of parts and is not open to sense-experience. But we can bring into awareness the reality behind appearances (unlike the theoretical reality of scientific posits). By emptying the mind of all conceptual and sensory content, a total inward stillness is produced. This state of imagelessness is not unconsciousness but instead permits the pouring in of a positive experience, an experience usually characterized negatively to set it apart from all other experiences. Nevertheless, the experience is felt to be an implosion of ultimate reality, accompanied by a sense of reality, certitude, and usually finality. The experience overcomes all sense of duality—of one reality set over against another—although this sense of nonduality is variously interpreted. The experience is not considered subjective or objective, that is, not based upon an individual subject nor independent of it. It is a contentless awareness, a pure light not illuminating any object but being its own content. There is no apprehension *of* unity, no *object* of awareness as in sense-experience and thought (for this would involve differentiation), but only the awareness, which itself is the reality.

Many aspects of mysticism could be discussed. But it is the alleged knowledge of the nature of reality given in mystical experiences that is of concern here, that is, a profound knowledge of reality enabling us to align our lives in accord with the way the world truly is. Knowledge in normal awareness is dualistic: we sense or think about something set apart from the experiencing subject. Mystics see this as involving a frame of reference that essentially distorts what is there. Instead, proper knowledge is given, not to one who thinks *about x*, but to one who *"becomes" x*. In the depth-mystical experience, all dualities are overcome; and in nature-mystical experiences, all sense of real, independent entities vanishes. In enlightenment, even beliefs become "internalized," not merely understood and acknowledged. In Theravāda Buddhism (S II.115), Narada is said to have the same knowledge as the enlightened Musila, but not himself to have achieved enlightenment; he understood and accepted the requisite truths, but had not experienced them. The analogy given is of a thirsty traveler who looks at water but does not drink—he understands but is not saved.

Both types of mystical experiences are normally of limited duration. The deeper kind could not last very long if we are to function in the world. And the nature-mystical experience usually fades quickly, with our normal awareness again taking over. Nature-mystical experiences may be extended by training, but as long as they remain special experiences they are temporary. But it is possible for an inner transformation of the total person—affecting cognitive and dispositional structures—to occur that integrates nature-mystical experiences into one's life constantly. Mystical "enlightenment" seems to involve internalizing a form of nature-mystical experience in this manner; it may also be an after-effect of the depth-mystical experience. The enlightened conduct themselves spontaneously because they have firmly internalized cognitive and emotive structures, making it unnecessary or even impossible to reflect upon alternative possibilities of conduct.[8]

The false world of distinct independent subjects and objects is seen through; the world of multiplicity (but not the reality behind it) is seen to be our creation. No energy is expended toward images. But sensory experience is occurring in the enlightened state (as part of the nature-mystical experience and after the depth-mystical experience). Stimuli *per se* are not illusory; only our misreading of their nature is. The mind, free from the disturbances our fabrications generate, is lucid and tranquil; thereby it reflects what is presented to it without adding or distorting anything.

Nature-mystical experiences involve an awareness of what is going on unconcerned with labeling what is presented to the senses. Concern is with the "beingness" of phenomena, not what categories we place them in (their being this or that). The "concentrated" mind of the mystic is neither attracted nor repelled by anything in particular; things are seen "as they really are." The total absence of concepts in sensory awareness occurs only in the extreme form of nature-mystical experience: in enlightened states, there is a structuring element reflecting the often complex classificatory schemes and soteriological goals of individual mystical systems. Concepts may not be the central concern, but discrimination and insight are integrated.

Does this *seeing* things correctly involve a difference in experience or only a difference in knowledge? Enlightenment does involve a permanent change in knowledge. Perhaps one has "transcended" dualistic experience only as one transcends a mirage by seeing that there is no water involved—one is no longer fooled by the experience but the experience remains the same. Ordinary awareness will have returned unaffected, but a new knowledge or significance will have been permanently added: one is no longer attached to surface distinctions. The alternative is that knowledge informs perception. A Gestalt-like switch occurs: the sensory stimuli remain the same but are structured in a new manner. Different facts then appear. For example, in an example from the Zen master Dōgen one sees the process of firewood's becoming ashes as a succession of different, causally unrelated entities (on analogy with the succession of the seasons) rather than as one entity undergoing change. As one previously *saw* causation, so now one does not see it. Physiological studies of meditation indicate a lack of habituation to the repetition of external stimulus.[9] This suggests a difference in experience, although it does not guarantee it. More important, the emphasis upon seeing correctly and clearly would indicate a difference in experience. That is, in enlightenment, the structuring element is permanently altered while the sensory stimuli remain the same. Whether the "state of consciousness" changes will be discussed later.

Mystical Experience in Context

From the recurrence of less ramified terms in the descriptions of mystical experiences, it seems correct to conclude, at least for the deeper variety, that mystical experiences are phenomenologically the same irrespective of the cultural or historical setting.[10] That is, the depth-mystical experience is the same regardless of the total conceptual construal of the experience. The assumption behind this position is that when anyone is similarly conditioned and all dualistic stimuli are removed, an experience occurs that is independent of culture and any conceptual scheme. Conceptual frameworks cannot affect the depth-experience itself (since the mind is emptied of anything conceptual), but would shape the understanding of it *after* the experience is over. In the case of nature-mystical experiences, concepts are absent only in the extreme instance; in the other instances, concepts may play an active role in the experiences themselves, thereby producing a variety of such experiences.

But even if the experiential element is identical in every instance, still mystical experiences occur only in the context of a specific tradition and era, and the understanding of them reflects the past experiences and beliefs of the individual mystic involved. No more than anyone else can mystics cut themselves off from their historical environment. There is no abstract "mysticism" but only specific mystical systems. Mystics' concepts, interpretations, and doctrines diverge. Their understanding of the experience and of the world often differs radically in fact. Some elements of a world-view are given in mystical experience—a sense of reality involving nonduality and a sense of great importance attached to this nonduality—but there is no conceptualization of the experience given along with this.[11] Śaṁkara construed the experienced nonduality in terms of the fundamental ontological basis of reality, while in Sāṁkhya-Yoga the nonduality is related to the *isolation (kaivalya)* of real individual subjects *(puruṣas)* from all matter *(prakṛti)*. No account, however, appears to be a minimal description of what occurs, impervious to error; each contains further belief-commitments. Each mystic feels that he or she sees reality as it truly is; from within a circle of belief, it is felt that one sees reality, and does not just give a theory-laden interpretation of it.

If mystics were interested only in inducing mystical experiences, if the experiences themselves were everything, doctrinal matters could be ignored as, in William James's term, "over-beliefs." But mystical experiences are usually embedded in conceptual schemes that place other concerns more centrally. Christian mystics place salvation through Christ and good works above mystical experiences. The Buddha was concerned with a radical end to suffering. But whatever concern is placed centrally, there still remains a need for placing the experiences in a context that gives them meaning and makes them intelligible to the experiencer. Difficulties arise in giving expression to the mystical insights (because of the otherness of the mystical and the experience), but the conceptual context is not valueless to the mystics. To dismiss the differences in understanding because of the common experiential component would be as unwarranted here as maintaining that any common experiential grounds between geocentric and heliocentric perceptions of the sun is sufficient to discredit the divergences between these points of view.

Only mystical experiences in the context of mystical systems give knowledge. What part the experiences themselves contribute to knowledge and what part the extramystical belief-claims contribute are hard to determine even if the experiential and conceptual components can indeed be totally separated. An additional complication is that there is a constant interaction between beliefs and experiences: the beliefs accepted at any point are shaped by previous experiences and vice versa. Mystical claims do not change so readily as scientific claims, however, because the experiential contribution remains fairly constant. In enlightenment, the mystic comes to *know* what was only understood before (although his or her understanding of the traditional teachings may be altered). The only exception to this is the mystic who attempts to devise a basically new system for understanding the mystical, a rare occurrence.

(2) Theravāda Buddhism

Theravāda Buddhism fits the pattern of a nature-mysticism and Advaita Vedānta of a depth-mysticism. Each tradition has a place for each type of experience: for Advaita, awareness of Īśvara appears to be a nature-mystical experience; and for the Theravādins, the depth-experience occurs in the trances *(jhānā)*. But each tradition values only one type of experience centrally: for Advaita, the awareness of the depth-mystical reality; and for the Theravādins, the permanent character-change of enlightenment (internalizing an insight into the nature of the flux of change).[12] This can be seen by looking at each tradition systematically. First the Theravāda tradition.

The Religious Problem: Suffering

The Buddha as portrayed in the Pāli *Nikāyas* is exclusively a religious teacher. The intention of his teaching *(dhamma)* is to effect a radical transformation within his hearers—to uproot one way of perceiving the world and to replace it with another. The religious nature of his program is generally affirmed by scholars.[13] The parable of the poisonous arrow reveals the practical bent of the teachings: when asked his opinion on some issues he considered speculative, the Buddha replied by comparing our situation to that of a person shot with an arrow; healing the wound, not knowledge about who shot him with what, is all that matters (M I.63). In our situation, "only what tends to aversion, the ending of grasping, cessation, calming, the paranormal powers, highest insight, and *nibbāna* is taught." How the Buddha dealt with the unanswered questions *(avyākata)* and many other instances point to the same emphasis: "I teach only two things: suffering and release from suffering" (M I.140).

Suffering (*dukkha:* dis-ease, dissatisfaction, frustration, or not getting what is wished for) is indeed the most fundamental concept in the Theravāda tradition. That all "constructed things" *(saṅkhārā)* are suffering (Dhp 278) is not to say that every experience is directly painful; in the heavens of the gods, in the trances *(jhānā),* and in sensory experience, there is pleasure. What is being emphasized is that in our present condition we are always open to suffering or loss of pleasure, and that constructed things, being impermanent, cannot ultimately satisfy us: even "from sensual pleasure sorrow springs."[14] Birth, sickness, old age, and death, and in fact every component constituting us are suffering (D II.305). This places us in a state of dissatisfaction, uneasiness, and insecurity.

The problem is compounded for the Buddha by his conviction of the reality of a cycle of rebirth *(saṃsāra).* A verse from the *Jātaka* tales brings out how much worse suffering now appears: "What misery to be born again / And have the flesh dissolve at death!"[15] Or from the Pāli canon:

> this round of existence is without known starting-point, and of beings who course and roll along from birth to birth, blinded by ignorance, and fettered by desire, there is no beginning discernable. Such is the length of time, O priest, during which misery and calamity have endured, and the cemeteries have been replenished; insomuch, O priest, that there is every reason to feel disgust and aversion for all the constituents of being, and to free oneself from them.[16]

The reality of rebirth shapes the religious quest of the Buddha: if there were no rebirth, the concern with suffering could be ended by suicide.[17] Instead he must search for "where no decay is ever known, nor death but all security," since "there is, there must be an escape! Impossible there should not be!" In light of this, consider what tradition has as the first words the Buddha spoke after his enlightenment:

> I have run through a course of many births looking for the maker of this dwelling and finding him not; painful is birth again and again. Now you

> are seen, O builder of the house; you will not build the house again. All your rafters are broken, your ridgepole is destroyed, [this] mind, set out on the attainment of *nibbāna,* has attained the extinction of desire. (Dhp 153–54)

The builder of the house (the cause of rebirth), according to the Buddha, is craving *(taṇhā).* By seeing things as they really are, all craving ceases and the cycle of rebirths dries up, and with the destruction of rebirth, the religious life is fulfilled: "Destroyed is birth; lived is the higher life; done is what ought to be done; and there is no future existence" (M I.184).

Dependent Origination

The mechanism by which suffering arises is elaborately set out in "dependent origination" *(paṭicca-samuppāda),* which in its fullest form contains twelve members. The arising of the aggregate of suffering is traced back through craving to the root cause, nescience (*avijjā,* "ignorance"). The basic nescience consists of four fundamentally misdirected views: taking what is by nature impermanent *(anicca)* as permanent, what can only bring suffering as pleasurable, what is repulsive as fair, and what is without self *(anattā)* as having a self (A IV.52). Nescience is not merely lack of correct understanding, but a positive misunderstanding of the nature of the experienced world. Suffering is generated by actions going against how reality actually is; its source is not the impermanence of reality *per se* but our nescience and attachment. All false views *(micchādiṭṭhī)* are based on the error that "there is" or "there is not" an eternally existing entity present that could be manipulated according to our desires.

Dependent upon the fundamental nescience, there arise dispositions or habit-energy *(saṅkhāra),* the residue from past misguided actions that shapes future action: our errors by repetition form habitual grooves for action. This habit-energy is connected to the Theravāda understanding of *kamma* (Skt., *karma*).[18] *Kamma* is the motivation *(cetanā)* behind acts done through the body, speech, and mind (A III.415). A kammic effect is not associated with every deed but only when greed *(lobha),* hatred *(dosa),* and delusion *(moha)* are operating. A deed in itself is kammically neutral; the motivation is what is central.

Habit-energy directs the discrimination *(viññāṇa)* that connects the karmic residue of past rebirths with a new embodiment *(nāma-rūpa)* upon which depend the six fields of sense-experience (the mind being considered a sense). Dependent upon these is sensation (*phassa:* "contact"), which forms the basis for pleasant, neutral, and painful feelings *(vedanā).* Feelings are the occasion for craving for sensory satisfaction *(kāma-taṇhā),* and for becoming or annihilation (M I.48, I.299). Dependent on these cravings, we grasp sense-pleasure and the theoretical supports for this quest (rules and rituals, views, and the self-doctrine) (D II.58, M I.67). The craving and subsequent attachment set in motion another instance of becoming *(bhava),* birth, and death. Thus another cycle of suffering rolls on.

The importance of this doctrine for the Buddhist tradition can be easily shown. For example, Assaji summarized the Buddha's teaching to Sāriputta by saying "The Buddha has told the conditions of all things arising dependently, and also how things cease to arise." Sāriputta, "seeing clearly and distinctly the teaching," immediately realized the "sorrowless state." Or more succinctly: "Who sees dependent origination, sees the teaching; who sees the teaching, sees dependent origination" (M I.190–91).

Dependent origination explains suffering without mention of permanence by showing how things arise conditionally; it is a formula of "this being so, that becomes." Conversely, on the *cessation* of a condition, what depends upon it does not arise. Therein lies its soteriological purpose: we have the ability to end suffering by removing the necessary conditions for the continuance of the cycle of rebirths; and dependent origination specifies what these conditions are. If the arising of suffering were the result of an unbreakable chain or if our actions had no predictable consequences, we would be powerless to effect our salvation. The doctrines of *kamma* and dependent origination show that this is not our situation. By removing craving, the cycle of suffering is stopped; by removing the fundamental nescience, craving is ended. Pleasant and painful feelings still occur in the lives of the enlightened (S II.82)—the Buddha himself experienced pain (A I.27)—but for them craving and attachment are no longer generated because of the change in their way of viewing the world. Consequently, their next death will be their last.

Right Views and Analysis

Therefore the only way to put an end to suffering is to put a final end to nescience and craving. This can only be accomplished by a complete transformation of the person. The Buddhist formulation of the means to this end is the four "noble truths." In form they parallel the ancient Indian practice concerning what to do about a *disease.* The Buddha is a metaphysician ending all suffering, as it were. In the first truth, the disease is diagnosed as suffering. Next, the cause is indicated (craving). Third, the cure (removal of craving) is to be directly experienced. And last, a treatment to accomplish this is prescribed: an eightfold "path" to enlightenment consisting of a code of conduct *(sīla),* concentration *(samādhi),* and insight *(paññā)* (M I.301). The various elements of this path to the rectification of our views and dispositions are to be cultivated together: how one leads one's life puts insight into practice, and so forth. For the purposes at hand, only three specific elements need to be discussed: right views, mindfulness, and concentration.

What are these right views, operating with which craving does not arise? Basically, they are the opposite of the beliefs revolving around abiding, distinct entities. They are characterized in many different ways (e.g., M I.486, III.71), but most simply they are seeing that every constructed entity leads to suffering, is impermanent, and is comprised of "elements of the experienced world" (the *dhammā*), which are without substance.

This analysis of the constructed entities *(saṅkhārā)*, which we normally think fill up our world as impermanent (Dhp 277), has a soteriological purpose, of course. In place of a world of independent objects, the Buddha proclaimed a continuous process of real, changing, conditionally arising components *(dhammā)*. All the physical and mental components of the experienced world—the sense-organs, the experiences themselves, and what is experienced (collectively the *dhātū*)—are composed of these factors. But although there is change and so nothing permanent, there is no total discontinuity between phenomena. Rather, there is a stream of successive states. It is a world-view denying the extremes that everything is permanent *(sabbam atthi ti)* and that nothing is real *(sabbam nātthi ti)* (S II.15f.). The factors making up the experienced world are not static but arise and pass away: the "being" who was is not identical to the "one" who is nor to the "one" who will be. Such factors constitute the reality behind appearances.

Two factors, space *(ākāsa)* and *nibbāna*, are not conditioned by other factors, but all factors are without self *(anattā)* (Dhp 279, A I.286). That is, everything is without any substance or unchanging physical essence isolating one entity from the rest. "Because it is void *(suñña)* of self or what belongs to self, it is said: 'The world is void' " (S IV.54). The most important of the illusory essences falling under this analysis is the self *(attan)*. In the place of a permanent "subject" underlying and substantiating individuality, a person is looked upon as made up of five material and mental aggregates of grasping *(upādānā-khandhā)*: form (*rūpa*, consisting of earth, water, fire, and air), discrimination or perception *(viññāṇa)*, ideas *(saññā)*, feelings *(vedanā)*, and dispositions *(saṅkhārā)*. These in turn are analyzed into many factors of the experienced world. The inner states exist, but are not changeless substances or agents. Like each aggregate, each act of perception arises and falls (M I.257). There is nothing permanent to grasp in the process; suffering arises from the error that there is. There is the perceiving without any substantial perceiver or object of perception. Perception occurs when the eye, eye-consciousness, and a form come together (S IV.86); the eye and the form are constantly changing and the eye-consciousness is dependent upon them. Every mental event arises and falls in the same way. Awareness endures only for the duration of individual acts (M I.257); there is no constant background awareness, according to such an analysis.

Each sentient being *(satta)* is a bundle *(kāya)* with no unchanging core. The idea of identity rather than continuity through change is only conceptual, a convenience for memory. The continuity of each nexus of forces is provided by the kammic residue of habit-energy. No organizing center, but an uninterrupted stream of ideas, feelings, and so forth shapes the future bundles of each "being." Death is the breaking up of one particular bundle of aggregates, but the stream continues undisrupted into another bundle by means of the discrimination shaped by habit-energy. The process of rebirth—or the transition from one slice to another within one rebirth—is likened in the *Milindapañha* to a constantly changed flame passing from one candle to another without an abiding entity being involved.

The religious import of this doctrine of analysis is to show how to look upon things without reference to anything permanent or personal. Thus if the nexus that "we" are is examined, nothing corresponding to "I" or "mine," no permanent subject thinking our thoughts, no being who "has" the experience, is found. Only in this way can we remove the root of suffering, nescience, which created the objects of craving. If everything is impermanent and without an abiding core, there is nothing to identify with and so nothing to satisfy, or anything to crave. All reference to an "I" other than the five aggregrates becomes unnecessary—only the action remains, and not an actor. Freed from a sense of "I," we are freed from the motivational greed, hatred, and delusion accompanying it. In fact, one is a "worthy" *(arahant),* that is, enlightened (S III.127–28).

Concepts of stable material objects and perceiving subjects become no more than names for temporary groups of factors of the experienced world. They are convenient constructions but ultimately only fictions overlaying the reality there. A modern analogy would be corporations as "legal fictions": there are assets and employees but no "thing" corresponding to the "corporation." The Theravādins would apply this idea to material objects and all of reality. "I" and "substance" become "mere sounds" or "practical designations": "Just as the word 'chariot' is used when the parts are put together properly, so is there the conventional usage 'being' when the aggregates are present" (S. I.135). There is only a flow of events; each "part" is a fabrication consisting in turn of smaller "parts." The ultimate realities are not subatomic smallest particles, but the *dhammā,* the factors involved in any experience. The listing of just what these factors are became quite elaborate in the Abhidamma phase of Buddhism; for example, the Theravāda Abhidhammists delineated 89 such factors. For soteriological reasons, the Prajñāpāramitā tradition reacted by emphasizing that the *dhammā* were void of substance. In either case, though, it is important to realize that what is being claimed is not that there is no reality "in" a "person," but only that our socially constructed self-image of individuality corresponds to nothing.

Meditation

Meditation (*bhāvanā:* "cultivation") trains the mind in the right view of things and the resulting dispositions. The untrained mind abstracts apparent permanence from experience and grasps these creations as "This is I" or "This is mine." Techniques of concentration *(samādhi)* and mindfulness *(sati)* cut through the projections of a self and material objects.

Concentration calms the mind by focusing attention upon a single point. The various lucid trances *(jhānā)* are gained by one-pointedness of mind *(cittekagotta).* Concentration is present any time awareness is focused, but the meditative practices cultivate it by progressively eliminating more and more sensory and mental content. The final result is an objectless awareness *(saññavedayitanirodha).* Outwardly a person in such a state appears dead; criteria were even listed (S IV.294) for differentiating a dead person from one in whom ideas and feelings had ceased. But these experiences are temporary—"constructed" by the mind—and consequently the fluxes *(āsavā)* of our ordinary, not fully concentrated mind that lead to craving are not destroyed. Thus the highest achievement of concentration does not accomplish the religious goal. *Nibbāna* involves an insight into the nature of the factors of the experienced world—something that occurs only outside the trances.[19] It is an "awakening" that cannot be forced by any exercise; and for this, the trances at best prepare that mind by setting up the necessary attention. Only a middle-level trance characterized by mindfulness and emotional even-mindedness (M I.21–22) is required.

What is needed for enlightenment is the other sort of meditation, mindfulness. It consists of bare attention *(sati)* to and clear seeing *(vipassanā)* of everyday occurrences and of special meditative objects (e.g., corpses in various degrees of decay for understanding the impermanence of our own bodies). Through mindful awareness of perception as impermanent, arising conditionally upon contact, and lasting only for the duration of the contact, one sees the nature of perception correctly. Other activities such as walking can be viewed in the same manner—from "I am moving" to seeing only that there is movement of the legs, the sensations, and feelings, but no "mover" nor one "body." When the body is mindfully viewed as a machine governed by a mental rope, such questions as "Who does the walking?" or "Whose is the walking?" do not arise. It is claimed that this is the *only way (ekāyano magga)* to abolish suffering. By applying this analysis to all our experiences, Buddhists claim that we overcome the delusory views and attachments.

Nibbāna

Insight *(paññā)* is simply seeing the true nature of constructed things (according to the Buddhist analysis). With this proper "knowing and seeing," the fluxes of the mind (sensory desire, views, nescience, and becoming) are necessarily destroyed (D II.230, M I.84). With the cognitive basis destroyed, the dispositions to act, which are as deeply rooted in us as the roots of a creeper in the ground (M I.43), are also destroyed. Thereby craving is ended; no effort is required because it is the nature of things *(dhammatā)* that a person who "knows and sees" becomes free of craving (A V.313). And simply, " . . . the extinction of greed, hatred, and delusion is what is termed *'nibbāna'* " (A I.38, S III.251). One began the path by the desire *(chanda)* for enlightenment replacing all other desires; but even this desire has been overcome: seeing correctly, one realizes that *nibbāna* is not a "thing" to be desired. A person in this state is free from any attachment or self-interest. *Nibbāna* is also the cessation of the discrimination involved in rebirth, and thus is the cessation of rebirth (S II.117). Thus suffering is ended, and this ends the religious problem with which the Buddha was concerned. A person who has accomplished this "has done what was to be done, has laid his burden down" (M I.4). There is nothing else to accomplish.

The nibbanized person has a reoriented point of view: once one has internalized the way of seeing informed by Buddhist insight, there is nothing to crave or to center one's life around.[20] Perceptions are like dreams: they occur but are not indicative of a plurality of real entities. Sensory-discriminations are structured according to the Buddhist analysis of the world. One has knowledge *(ñāṇa)* and the seeing *(dassana)* of things as they really are *(yathābhūtaṃ)* (A III.200). One sees the factors of the experienced world as they really are, and therefore the illusory potential of sense-experience is forever removed. There is no twisting of reality to conform to hopes based upon the delusory sense of a "self."

How concepts are seen is also changed. Concepts like "self" and "being" become "conventional usages *(sammuti),* convenient language, worldly terms of communication, conventional descriptions by which the Thus-gone one *(tathāgata)* communicates without misapprehending them" (D I.195). Concepts *per se* are not distortive but useful, and the enlightened person makes use of conventional forms of speech without being misled by them (M I.500). Following the chariot-analysis, the enlightened can use "I" or any other term designating entities without positing an enduring, unchanging referent. The tendency to take our concepts as indicative of the nature of reality is known as "conceptual projection" *(papañca).* From sense-contact, feelings arise; what one feels, one forms an idea of *(sañjanāti);* what one forms an idea of, one reflects upon *(vitakketi);* what one reflects on, one projects *(papañceti);* and subsequently one is assailed by these projected concepts (M I.111–12). The enlightened make conceptual and perceptual distinctions, and reason with concepts, but they do not get caught up in the inventions of the mind by which the unenlightened create a world of multiple entities. The enlightened know the nature of a "person" or *"nibbāna."* They know that these are not "things" that can exist or be possessed.

Concepts in their proper place are useful, but once they are removed from that situation, they only lead to confusion. Thus the Buddha left such questions as whether the Thus-gone one exists, does not exist, both exists and does not exist, or neither exists nor not exists after his death unanswered *(avyākata)*, because any answer does not fit *(upeti)* the case, and would only supply an image for attachment rather than lead to enlightenment (M I.431). The "self" is not destroyed because there is no self to destroy. Once the enlightened have got rid of the five aggregates, "arise" and "does not arise" do not fit (M II.166). The enlightened cannot be spoken of because they are without the factors of the experienced world (*Sutta-nipāṭa* 1076). It is only "by not knowing, by not seeing" the aggregates properly that one holds a view concerning whether the enlightened exist or not after death (S IV. 373–402). All the Buddha would say about what occurs after his death is: "On the dissolution of the body, after the end of his life, neither the gods nor men will see him" (D I.1).

The Status of Buddhist Claims

The Buddha's reluctance to discuss certain matters, and the religious nature of his doctrine have led some to doubt that the Buddha made any knowledge-claims about the nature of reality. His teachings are similar to a raft for crossing over to the "other shore"; they are merely an expedient not to be got hold of but abandoned once their function is fulfilled (M I.22). Thus when the *Sutta-nipāṭa* (1119) says "Consider the whole world as void *(suñña)*," supposedly nothing about the real nature of the world is being claimed, but only a technique for obtaining enlightenment is being recommended. Dependent origination, the *dhamma*-analysis, and the central concepts of suffering and *nibbāna* all supposedly make no claims outside the religious framework. That the Buddha taught contradictory doctrines to different listeners according to his knowledge of their capacity to understand, for example, teaching the laity that there is no mental development while teaching the disciples that there is (A I.10), is taken to indicate a lack of interest in matters of "truth."

In other instances, truth definitely figures in the claims about the world. Suffering might involve only our subjective reactions, and contain no reference to the world. But the other two marks (impermanence and lack of substance) are hard to construe as other than claims about the nature of the world. The constituents of experienced reality—the *dhammā*—include not only sense-experiences but also the sense-organs and sense-objects. The *dhammā* are not our subjective creations, nor do the Theravādins equate all of reality with experience. They are "realists" in that sense. Questions related to the temporal and spatial extent of the world, the relation of the individual *(jīva)* to the body *(sarīra)*, and so forth, were dismissed unanswered as positive impediments to the religious quest. But other aspects of the world as experienced are relevant. Whether there is a cycle of rebirths and whether the enlightened have escaped from it are matters of truth and of utmost importance to this religious system. Wrong views mentioned often are not believing in rebirth, in the consequences of good and bad deeds, and in this world and another (D II.316–17, M III.71). If these are not accepted at face value, suicide would be a successful course of action for the unenlightened to end suffering.

When the Theravādins claim "Suffering exists" (S II.19) or that the voidness of the factors of experience is true whether the Buddhas exist or not to point it out (A I.285), these beliefs should be accepted as the factual component of a religious way of life, not advice given regardless of the nature of the world. Their belief-claims are not just techniques to bring about a desired end. Sometimes it is claimed that the Buddha taught the no-self doctrine only because people are attached to a sense of permanence—had people been attached to impermanence, he would have taught the doctrine of self. But people have a sense of order and yet the Buddha did not teach a doctrine of the randomness or disorder of events. Something besides mere expediency is involved. Buddhists are convinced that they see reality correctly, that they see things "as they really are" *(yathā-bhūtaṃ).*

But again this does not mean that the Buddha had an intellectual interest in the world outside the religious enterprise. The practical force of his interests is revealed in the parable of the handful of leaves (S V.437): if one compares the number of leaves the Buddha had in his hand to the number of leaves in the forest, the immense difference is obvious. "Even so, monks, I have told you only a little of what I know; what I have not told you is much greater. And why have I not told you? Because that is not useful, . . . not leading to *nibbāna.*" The Buddha was pragmatic in the sense of being concerned only with what was conducive to enlightenment, but not pragmatic in the technical philosophical sense of equating truth with what works; he did not express some things, not because they were neither true nor false, but only because they were not helpful to the religious problem at hand. So too his "skillful means" in adjusting his teaching to the capacity of the listener does not preclude that the answer given the advanced disciples is correct. His final word on the world is as it appears within this framework, namely how the world must be composed for our experiences relevant to overcoming suffering to occur as they do. Such claims may not be about the ultimate status of the world apart from our experiences and concerns, but they are about the world nonetheless. Succinctly, they state the nature of the world *(dhammatā, tathatā).* The Buddha is interested only in what works, but it works because it is in keeping with reality.

This silence on ultimate matters has been construed by some as an endorsement of an ultimate reality "behind" the *dhammā.* The Buddha discussed what goes on within the dream without dealing with the complete status of the dream and dreamer, as it were. For instance, it is claimed that the Buddha denied that there is any self to be experienced in the aggregates or apart from them, but not that there is absolutely no self. George Grimm, for one, interprets the Buddha's teachings on this point as seeking the self indirectly by removing all that is not the self: "You teach the *Atta,* but I teach what the *Ātta* is not."[21] So also the remarks concerning the enlightened after death would suggest *prima facie* some sort of existence (e.g., a "worthy" released from form is immeasurable or hard to fathom like a great ocean); if there were no existence at all, it could be so stated. And similarly regarding *nibbāna:* there are passages in the Pāli canon that give *nibbāna* an ontological status greater than the state of a person in whom the fires of greed, hatred, and delusion have abated. *Nibbāna* is equated with the "invisible infinite consciousness which shines everywhere" or a place or entity identical with absolute reality.[22] *Udāna* 80 is often cited as maintaining that *nibbāna* is absolute reality: there is the not-born, the not-become, the not-made, the nonconstructed, otherwise there could be no escape from the born, become, made, and constructed. This could refer only to the *state* of *nibbāna.* But in any case, the Buddha as most usually portrayed in the Theravāda canon did reject any views to which we may become attached, and any views concerning an ultimate reality would be within this category. That something positive was actually intended will remain a matter of speculation.

(3) Advaita Vedānta

Background: Buddhism and the Upaniṣads

Turning to Śaṁkara's Advaita, the major difference at least in emphasis is obvious: while the Buddha was averse to discussing the ultimate status of the world, Śaṁkara revels in it. There are lines of continuity, however, between the two traditions. Both share something of a common religious framework: the roots of Śaṁkara's thought lie in the *Upaniṣads,* which arose in the cultural environment that also produced the Buddha. In addition, Śaṁkara was indirectly influenced by Mahāyāna Buddhism through his teacher's teacher, who probably began his career as a Buddhist. There are some Buddhist ideas in his commentary on the *Brahma-sūtras:* he mentions that the desire for release results from the reflection that all effects, objects, and powers are impermanent (I.3.26); and the commentary concludes with mention of *nirvāṇa* (IV.4.22). The *Viveka-cūḍāmaṇi,* which Śaṁkara may have composed, contains many Buddhist images: becoming calm like a flame with fire consumed, the need to cross the ocean of this world (with its shark of craving), the wheel of birth and death, *nirvāṇa,* and ignorance as the cause of rebirth. The importance of knowledge in the Theravāda ("he who knows and sees") and the *Upaniṣads* ("he who knows thus") certainly carries through to Śaṁkara. But the differences are at least as important. Central to this is the difference in at least soteriological technique just mentioned: Śaṁkara's willingness to discuss the relationship of the world of becoming to the still depth-world of being in contrast to the Buddha's silence on anything concerning the latter. He was not more speculative than the Buddha: each discusses experiences and sensory matters but interprets them differently. They differ in that Śaṁkara gave the depth-mystical experience extreme importance while the Buddha valued a state of mindfulness more.

Śaṁkara's chief opponents in his *Brahma-sūtra-bhāṣya* are the Sāṁkhyins, but some criticism is leveled against the Sarvāstivādins, Mādhyamikins, and Yogācārins—the Theravādins at the time no longer existed in India. Of concern to the Theravādins, Śaṁkara attacked the theory of momentariness *(kṣanika-vāda)* held by the Buddhist realists, although it is not present in the Pāli *Nikāyas:* how can the self be momentary (i.e., arise and fall completely every moment) if memory occurs (BSB II.2.20)? So also he claimed that Buddhists cannot explain how the aggregates are brought together: how do the aggregates congeal to form an individual unit (BSB II.2.18)?

It is upon the *Upaniṣads* that Śaṁkara built his system.[23] These texts evolved out of Vedic speculation tending toward a monism of powers behind phenomena (Brahman, originally the power of the ritual utterance, being eventually deemed that power) and a concern with the self; the exact point in this evolution of thought at which mysticism entered is not clear. The *Chāndogya* is most often cited by Śaṁkara, but it and the *Bṛhadāraṇyaka* hold fundamental importance. Both assert that the self and Brahman are identical (e.g., BU II.5.1; CU III.14), although Śaṁkara and other Vedāntins such as Rāmānuja differ on the meaning of this. The *Chāndogya* focuses primarily on the sensed realm, as in the passage concerning knowing the clayness of all clay objects (VI.1.4). The *Bṛhadāraṇyaka,* on the other hand, deals more with the "unseen seer" and how the self is the source of the sensed objects (II.1.20). One of its key sayings concerns the "inner controller" who is "not this, not this" *(na iti na iti)* (III.9.26, II.3.6, IV.2.4, IV.4.22, IV.5.15). It discusses two forms *(rūpas)* of Brahman (II.3), only one of which can be understood: for how can one understand the inner controller, the self who understands? How can one know the knower? The *Bṛhadāraṇyaka* deals more with negatives and apparent dualities, while the *Chāndogya* emphasizes only the reality *(sat)* of Brahman.

Each of these *Upaniṣads* and the corpus of the *Upaniṣads* as a whole are not uniform in their doctrine. Śaṁkara needed to interpret them to fit them into his system. He had to treat some passages as figurative in order to achieve a consistent whole. Thus, for example, the *Bṛhadāraṇyaka's* assertion that nonbeing alone was in the beginning is interpreted to mean that everything arose out of "being without name and form" (BSB I.4.15, II.1.17). Passages in the *Bṛhadāraṇyaka* (I.4.1) and the *Chāndogya* (VI.2.1) giving *sat* precedence are favored by being interpreted as literally true. Śaṁkara in addition had trouble with the nonrevealed texts. the *Bhagavad-gītā* with its theistic thrust presents major difficulties for him. Even Bādārayaṇa's *Brahma-sūtras* may be more supportive of Rāmānuja's position than Śaṁkara's.

Brahman / Ātman

In attempting to present systematically Śaṁkara's ideas, one problem immediately presents itself: all the categories we differentiate (ontological, psychological, epistemological, soteriological) are conflated into one. Brahman is the act of knowledge, the reality known, the knowledge itself, the knowing subject, and liberation. But despite this, various strands of ideas within Advaita Vedānta can be isolated to a degree.

The most useful analogy for understanding Advaita is that of a dreamer and a dream: the dreamer is the self *(ātman)*, which is the only reality; Brahman is whatever reality there is in the dream (i.e., the dream-consciousness); the principal subject within the dream and any other characters are the individuals *(jīvas)*; and the dream-phenomena in general are the realm of nescience. Ultimately, there is only one reality constituting all the dream-phenomena and that reality is consciousness (the dreamer): the ground of the dream (Brahman) is the ground of the dreamer *(ātman)*. The consciousness of each character in the dream is absolutely identical to that consciousness sustaining the dream, although this is not apparent to the characters in the dream who have not awakened.

Brahman is the unchanging "ground," the "inner self," of all changing phenomena. It is the conscious all-pervading "ground" from which proceed the origin, subsistence, and dissolution of the world (BSB I.1.2); it is the "cause" of whatever is (BSB II.3.35). "Ground" and "cause" are placed within quotation marks because in fact Brahman is the sum and substance of any reality: there is no reality *(satya)* but Brahman. Any further specification of the nature of Brahman is rendered difficult by the fact that it is considered not to be an *object* among objects or a thing with parts. Whatever is perceivable by the senses may be indicated by type, quality, function, and relationship; but since Brahman cannot be sensed and does not possess any differentiation, we cannot explain its nature (KUB I.4). It is not the factors of the experienced world void of substance of Buddhism nor the undifferentiated being of much other nature-mysticism, for this would make Brahman a perceivable or objective whole with parts. Rather it is the timeless, spaceless ground out of which all sensed reality arises. Śaṁkara uses "nondualism" *(a-dvaita)* to refer to his system rather than monism (perhaps *"ekaita"* in Sanskrit) in order to remove the possibility of misinterpretation in this regard. Nondualism is intended to deny any monism or pluralism, for, if Brahman is not an object open to discrimination, then no number—including "one"—applies. Any noun or pronoun—even "it"—would imply that Brahman is an object of some sort, and so all are denied. Therefore Brahman is "not this, not this," that is, the negation of the reality of any "forms" fictitiously attributed to Brahman becomes central. By eliminating all characterization, all that remains is the pure Brahman, the reality of the real *(satyasya satya).*

It is utterly impossible to describe Brahman's nature free from the differences due to limiting adjuncts (BUB II.3.6). Even "being" or "nonbeing," "is" or "is not" do not apply (BGB XIII.12). Śaṁkara distinguishes two forms *(rūpas)* of Brahman: Brahman as an object of knowledge and as an object of nescience. Brahman in the higher form, that is, free of all qualifications *(nir-guṇa),* can only be designated as "not this, not this" (the elimination of all attributes). When this form of Brahman has properties attributed to it, it is only for the purpose of directing our mind toward it (BSB III.3.12). But this distinction is made from within the realm of nescience: Brahman as it really is is beyond both forms (BUB II.3.1, BSB III.2.22). Even as an object of *knowledge,* Brahman becomes an object, a conception by a character in the dream. Ultimately, even the words "Brahman" and "self" are merely superimpositions upon reality (BUB II.3.6); even the concept "reality" would be a problem.

Some positive characterizations are more applicable than their negatives for directing the mind to Brahman. But problems still arise. One such characterization occurs in the *Taittirīya Upaniṣad* (II.1.1): Brahman is reality *(satya),* knowledge *(jñāna),* and infinity *(ananta).* According to Śaṁkara's commentary, these are meant to define Brahman, but each functions to remove some attribute, including the other two mentioned. Thus: "If Brahman is the agent of knowing, reality and infinity cannot truly be applied to it. For as the agent of knowing, it changes; and as such, how can it be real and infinite?" "Knowledge" negates materiality, and the other two attributes negate knowledge. Thus a process of negation is still involved.

Brahman within the context of a person is the self *(ātman).* For the *Upaniṣads,* the self is the ear of the ear, the mind of the mind, and so forth, that is, the inner knower and controller that dwells in the "space in the heart." It is the "essence" of whatever is being discussed, the reality of the real. It is the awareness present in every finite act of sensing, thinking, and self-awareness. This awareness never changes: it is undecaying, immortal, uniform, consisting of understanding *(vijñāna-ghana)* only, infinite, continuous (BU II.4.12). For Advaita, the self is not active, but instead an "inactive controller," all actions being parts of the realm of nescience. Awareness is not an organizing activity of the brain, nor is it ever an object of itself. Rather for Śaṁkara awareness is like a self-generating, colorless beam of light that illuminates, but cannot reflect back upon itself—it is never an object *within* awareness. What we construe as objects is merely the spectrum produced by the prism of nescience, not the source, which alone is real.

The unchanging "knower" cannot be known by itself any more than fire can be consumed by the consuming flame (KUB II.1).

Although, like the Buddha, Śaṁkara did not believe in anything substantiating an abiding individual *(jīva),* he did believe that there is a depth-reality that can be denoted by "the self" *(ātman)* if any term is appropriate. That is, for Śaṁkara, "I" has a meaning outside of the context of individual existence. The individual and the real self differ only in name, since the latter constitutes whatever reality the former has (as well as the reality of all other phenomena). Both the Buddha and Śaṁkara would agree that there is no unchanging essence in the *content* of our awareness that is identifiable with a self. The Buddha, somewhat like Hume, affirmed that a self was not perceived, and thus that the concept "self" was only a convenient term for a "bundle of perceptions." For Śaṁkara, construing consciousness as something that comes and goes or varies with each act would be a misidentification of the self with the impure content of awareness.

The sense of "me" (the experiencing subject objectified) is the paradigmatic instance of nescience: there is no inner agent but only an awareness that cannot be objectified or limited to a body. The awareness is the only reality, not what is experienced, just as in dreams: dreaming itself is real, but the content (the characters and events within the dream) is not. When the self has the limiting adjuncts of the body, it is the individual *(jīva)* of the cycle of rebirth (BUB III.8.12). The self "enters" the "city of Brahman" (the body), but the only reality involved is actually the self—the idea of a body or the individual as a real entity is nescience. There is a *prima facie* dualism here: the body and the awareness existing independently of the body. Śaṁkara does discuss the relation of consciousness to the body: consciousness cannot be a product of the body (for then the body would know itself), consciousness needs the body without being a product of it (just as sight needs a lamp without being a product of it), and so forth (BSB III.3.54). Charles Tart's distinction between a fundamental *awareness* that may not be a function of the brain (i.e., exist independent of it) and *consciousness* modulated by the structure of the brain is appropriate here also, although Tart remains a dualist.[24]

But Śaṁkara ultimately is a nondualist: "mind" and "body" are the same stuff (awareness). For Advaita, the unchanging awareness that constitutes our true self is absolutely *identical* with the awareness within every person, and also with the ground of the world (Brahman). Everything is ultimately nondifferent *(na apara)* from the tranquil inner controller: my self, which is smaller than a grain of millet, is greater than all the world (CU III.13.7). We are not composed of the same substance, nor are we parts of one great whole, but we are identically the same reality. As Brahman is not an object, it is extensionless, consists of no parts, and is in every way nondual; all distinctions operate only in the realm of dualistic awareness, the realm where reality is not revealed. The doctrine of difference-and-non-difference (*bhedābhedavāda,* on the analogy to sparks being both different and not different from fire), and all doctrines of ultimate dualities and pluralities are rejected. Thus, what we normally consider *objective* "does not fall within the category of 'object' and constitutes the inner self *(pratyag-ātman)* of all" (BSB III.2.22). Therefore when the *Chāndogya Upaniṣad* (VI.8.7) says "That is the real *(satya);* that is the self; you are that *(tat tvam asi),*" Śaṁkara interpreted it as meaning that *"tat"* denoted Brahman (the reality in objects), *"tvam"* the inner self, and the verbal form of "to be," *"asi,"* absolute identity. Other Upaniṣadic "great sayings" for Advaita are "I am all this" (CU VII.25.1), "All this is the self" (CU VII.25.2), "I am Brahman" (BU I.4.10), "All this is Brahman" (*Muṇḍaka Up.* II.2.12), and "This self is Brahman" (BU II.5.19). More properly, since there is no object involved, we cannot speak of a *subject* either, as Śaṁkara himself confirmed (TUB II.1.1). The permanent, unchanging awareness itself is the one ultimate reality.

Is Advaita realistic or idealistic in this view of reality? Within the realm of everyday experience, Śaṁkara argues against the idealist view that there is no external reality (BSB II.2.28). Yet ultimately the self constitutes the object of sight (BSB I.3.13). So also he argues against an unconscious material (the *prakṛti* of the Sāṁkhya) as the basis of the world: consciousness is necessary for the orderly arrangement of the world (BSB I.1.15). Nor is his position one of nihilism: the "not this, not this" analysis does not result in total nonreality; one can deny something as unreal only with regard to something real (here, Brahman) (BSB III.2.22). But his nonobjectivism should not be construed as solipsism. The self is a reality encompassing more than the subjective personal consciousness. To use the dream analogy: it is not a character within the dream who is the source of the dream's reality. That the common source of all the dream-phenomena (the dreamer) is seen as a character in the dream does not change this. The whole universe is not contained within the *jīva,* the individual character. That reality *(satya)* is not material does not obviate the realistic nature of Advaita.

The Status of the World

The world of appearances, the realm of apparent change, is from the point of view of nescience, an "emanation" from Brahman. The phenomenal world is the "product" of Brahman seen as Īśvara, that is, Brahman with qualifications *(sa-guṇa),* through his creative power *(māyā).* The omniscient, omnipotent Īśvara creates, sustains, and destroys the world as mere sport *(līlā),* which proceeds from his own nature: no more purpose is involved than in our breathing in and out (BSB II.1.33). Since Brahman is the sum of reality, Īśvara is the formless Brahman misread through a degree of nescience. In this form, Brahman is an object of devotion, and is experienceable through a nature-mystical experience, but this is still subordinate to the higher form and the deeper experience.

"*Māyā*" originally referred to the creative power *(śakti)* of the god Varuṇa, a power deceiving people. It took on the connotations of a magician's power to deceive in an "illusion." It is the dependence of phenomena upon the magician and the deceptiveness of the illusion taken at face value that lie behind the Advaitin use of the term. Like many concepts in Advaita, "*māyā*" is multidimensional. It is the cause of the sensory realm, ontologically the "substance" of it and epistemologically the active falsification of knowledge. In all these regards, *māyā* is equivalent to nescience *(avidyā)*, that is, seeing a plurality of real things. We take the colored spectrum to be constituted of individual real things and to be the sum of the reality involved.[25] Our concepts come to be taken as describing the sum of reality in this positive misreading of the nature of Brahman.

To understand Advaita's notion of nescience, superimposition *(adhyāsa)* must be understood. Śaṁkara defines "superimposition" as "the apparent presentation, in the form of remembrance, to consciousness of something previously observed, in some other thing."[26] Such projection produces an illusion. Two types of superimposition are distinguished: (1) The mundane variety *(prātibhāṣika)* such as mother of pearl appearing to be silver, and (2) the root-nescience *(mūlāvidyā)*, the superimposition of objects and their attributes upon the self that is not itself an object. The self is subject to superimposition only because it is open to "experience" (i.e., can be realized in the depth-mystical experience). We superimpose onto the self the body, sense-organs, the mind, and extrapersonal attributes; and the inner self is superimposed onto these. We misidentify ourself with the mind, body, and so forth, and fail to discern the identity of the self and Brahman. We couple the real and unreal in such expressions as "That am I" and "That is mine." Similarly, the idea of Brahman as an entity is superimposed upon the name "Brahman" (BSB III.3.9). In short, we superimpose dualistic features on the self and the self on what is not the self (BSB III.3.9). Mundane superimposition can be corrected by later sensory experience (i.e., within the context of the "dream," some perceptions are accurate and some not). But the root-nescience can be corrected only by the liberating knowledge *(vidyā)*, the discernment of what is real and what is projected.

The world cannot be declared to be real (*sat*, applicable to Brahman only) or unreal (*asat*, like the hare's horn or the son of a barren woman) (BSB II.1.14). Instead, *māyā* is a middle indefinable *(anirvacanīya)* realm: the experienced world is not totally unreal (because it has a real "basis") nor real (since we misconceive it). Like a dream (although the world is more durable than a dream [BSB III.2.3–4]), there is something there, but we normally misread its status; and it is not equal to the dreamer.

In the *Upaniṣads*, Brahman is the root of the world (*Kaṭha Up.* II.3.1):

"as a spider moves along the thread, as sparks arise from the fire, even so from this self arise all breaths, all worlds, all gods, and all beings; its secret meaning *(upaniṣad)* is the reality of the real *(satyasya satya)*" (BU II.1.20). For Śaṁkara, the world is merely superimposed upon reality: it does not really "emanate" from Brahman as the result of a transformation. Distinctions (the basic subject-object duality and the separation of distinct subjects and objects) are the result of nescience: from the standpoint of the highest truth, the self is one but it appears multiple due to nescience, just as the one moon is not really multiplied by appearing double to a person with defective vision (BSB II.1.27; *Upadeśasāhasrī* I.40; cf. TUB II.4.1). Śaṁkara used *Chāndogya Upaniṣad* VI.1.4 as an illustration: "Just as by one clod of clay all that is made of clay is understood, the modifications being only a name arising from speech while the reality is just clay." His comment: "Insofar as they (the distinctions) are names, they are unreal; insofar as they are clay, they are real *(sat)*." On the surface all objects are not identical, but only in the reality of their depth. Thus golden ornaments and figures are not identical except insofar as gold constitutes the self of each (BSB IV.1.4). Once we start to differentiate reality into distinct entities according to "name and form" *(nāma-rūpa)*, we enter the realm of nescience. Language plays a crucial role: since nescience is the distinctions made by names and forms, it arises from speech alone (BSB II.1.27). So also with Brahman: Brahman is unreal as far as it is distinguished as an entity by its name.

The relation of Brahman to the world is that there is no relation because Brahman alone is real—Brahman (the dreamer) appears to be diverse (the dream) because of our mis-seeing (in the dream), not because the changeless Brahman can be multiplied or transformed in any way. The relation is not exactly that of "appearance" to "reality," since the reality can be directly realized. In one sense, we really "see" only Brahman all the time: in whatever experience is occurring, it is the only reality involved. We *always* "see" it—that is why it is considered real (permanence being the criterion of reality). But nescience-guided perception misreads what is experienced—like *seeing* a rope, but taking it to be a snake. It is the "effects" of Brahman, not Brahman itself, that is directly sensed in normal experience (BSB II.1.27). Thus Brahman constitutes the reality of what is seen. In realizing Brahman without qualifications, we are totally free of error; in experiencing qualified Brahman we misread Brahman in taking it to be an object; and in normal experience a total misreading occurs: the errors have totally replaced the reality.

The principal difficulty in maintaining absolute nonduality involves the status of the limitations of name and form that constitute nescience. Is the realm of nescience a reality too?[27] Śaṁkara maintain that Brahman is the sole source of reality: no other reality is required in any way. Śaṁkara discusses phenomena as if they were things limiting Brahman in one way or another. For Advaita, the reality in the limitations is Brahman too: Brahman is the light refracted through the prism of nescience into a spectrum, but the prism too is the result of the light itself. All the analogies Advaita appeals to involve dualities that give at least some reality to the dream-realm. None account for how the realm of change and apparent duality arose. And if the dream-realm is eternal (although in flux), as Śaṁkara admits, and if the content of waking consciousness is never negated (BSB II.2.29), then the realm of change (if not its content) fulfills the criterion of permanence too. Rāmānuja then seems correct in asking how accepting the realm of change as neither real nor unreal but as "indefinable" differs from accepting it as real: what difference is there in saying the realm of change is not totally unreal and saying it is real? The dependence of this realm on another reality—the dream upon the dreamer—may be emphasized, but this does not overcome the basic problem.

Knowledge and Freedom

Our basic nescience, according to Śaṁkara, as with the Theravādins, cannot be replaced by a minor adjustment of our views of the world, but only by a total switch. Discernment of the self is the requisite knowledge *(jñāna, parāvidyā, brahmavidyā).* All knowledge is practical to the extent that it directly affects our lives. But knowledge of Brahman transforms our lives totally in that it effects (and *is*) release *(mokṣa)* from the cycle of rebirths. It is final, and only this knowledge can result in immortality (BSB I.4.24).

In the *Upaniṣads* too, knowledge is of utmost value. It is a transformative *power* since "who knows thus" *(ya evaṃ veda)* becomes *(bhavati):* who knows *x* obtains or becomes *x*. This is true in the matter of fulfilling mundane desires (as in the *Chāndogya*) and most vitally in obtaining release from rebirth. Who knows Brahman becomes Brahman or the whole of reality (BU IV.4.13,17,25). One knows everything in the sense that one knows the *source:* omniscience of the surface level is not given. One does not know each possible clay object, but only the clayness of each. Nor is *control* given; depth-enlightenment does not necessarily confer paranormal powers helpful in controlling the universe and one's fate.[28]

In deep sleep and after death, we automatically "merge" with Brahman, that is, nescience is stilled and we sense no separateness even if we still have no awareness of the identity present always.[29] In actuality, we are Brahman at all times, but are unaware of it because of the covering of nescience. But when we *know* Brahman and "become" Brahman in enlightenment, we are out of the cycle of rebirths (BSB I.1.14). The "subtle body" involved with karmic retribution is dissolved at release, not death (BS IV.1.13–14), although the effects of old deeds must come to fruition before the final merging can occur. For the enlightened who bring knowledge to this event, no new body can be produced since there is no craving generated by a sense of individuality to produce a personal "emanation."[30]

The *Muṇḍaka Upaniṣad* (I.1.4–6) distinguishes two types of knowledge: higher *(parāvidyā)* and lower *(aparāvidyā)*. The latter is knowledge of the Vedas and its auxiliary texts and fields of study, while the higher is knowledge of the ungraspable, undecaying source of all beings. In his commentary upon the passage, Śaṁkara declared that ultimately the lower knowledge is nescience, which has to be eradicated since nothing of reality is known by knowing the object of nescience. Within the realm of nescience, there are various authorities *(pramāṇas)*. Śaṁkara relied heavily upon testimony *(śabda)* of the revealed texts (*śruti*, that is, the Vedas) and discussed in addition only sense-perception *(pratyakṣa)* and inference *(anumaṇa)* in the *Brahma-sūtra-bhāṣya*. These authorities have for their object something dependent upon nescience. And each of them is supreme in its area: testimony is absolute in the understanding of Brahman—reason alone is incapable of demonstrating the nature of reality, as all the contradictory theories reveal (BSB II.1.1). But testimony is not authoritative outside this area: a hundred scriptural texts declaring fire to be cold or without light would not make it so (BGB XVI.66).

Even the Brahman of the Vedic texts is dependent upon nescience (BSB, introduction). Within the realm of nescience, Brahman is known based upon the authority of the Vedas and as being the inner witness of persons. Since it is not an object of the senses, it has no connection with the other authorities (BSB I.1.2). The depth-mystical experience or any other special experience is *not* considered authoritative for Śaṁkara because we still need to understand it and the Vedas provide the proper interpretation. The Vedas are deemed authoritative among orthodox Indians because (it is believed) they arose with the origin of the world—the seers *(ṛṣis)* merely saw them, and did not compose them.

In addition to the testimony of the scriptures, the existence of Brahman is known on the ground that it is the self of everyone (BSB I.1.1). Śaṁkara went as far as to say that it is impossible to claim that the self is not apprehended (BSB I.1.4). The self cannot be denied, for who would the denier be (BSB I.1.17)? But as long as we are within the realm of nescience, this argument is not self-evident. The normal sensing subject is based upon the subject-object duality: I know and sense things distinct from myself. Likewise when I say, following Descartes's method (cf. BSB I.1.17, IV.1.3), that I *know* I exist, I am aware only of an individual existence—that my "I" is the same as every "I" (i.e., that there is only one *ātman*) and that this "I" is the ground of "objective" reality (i.e., Brahman) is not something I am aware of. Śaṁkara conceded that there is a conflict of opinions concerning the nature of the self; hence the need for the *Brahma-sūtras*. And therefore the revealed texts once again become central.

The knowledge of Brahman contained in the Vedas culminates in the realization of Brahman (BSB I.1.2). Higher knowledge, said Śaṁkara, is not the mere assemblage of words in the *Upaniṣads* but is the knowledge these texts impart (MUB I.1.5–6). It is the required direct awareness (*samyag-darśana,* "right seeing") that results from the inquiry into Brahman (BSB I.1.2).

Meditative concentration is essential for this knowledge to occur. The later *Upaniṣads* deal explicitly with *yoga* (meditative discipline or practice), for example, *Śvetāśvatara* I.3, II.8–9. The *Maitrī* (VI.18) speaks of a sixfold *yoga* including breath-control, withdrawal of the senses, lucid trance *(dhyāna),* and concentration *(samādhi).* The *Kaṭha* deals extensively with *yoga* and related issues: the self cannot be grasped by the senses (II.1.1): one needs to be tranquil, with a concentrated mind (I.2.24); the senses must be turned inward (II.1.1); *yoga* is the steady control of the senses (II.3.11); by the mind alone can the seer be attained (II.1.11); one cannot see diversity (II.1.10); the mind and senses must be tranquil (I.2.20), not stirring at all (II.3.1). Both the *Kaṭha* (I.2.23) and the *Muṇḍaka* (III.2.3) say enlightenment is attainable only to those whom the self chooses (i.e., no practices can force the experience). Even the early *Upaniṣads,* texts of the transitional period from Vedic speculation to this later mysticism, contain a few such passages. The *Bṛhadāraṇyaka* says the self is not open to sense-experience.[31] So also we need to be calm, self-controlled, withdrawn, patient, and collected to see the self (IV.4.23). The *Chāndogya* too speaks of the necessity of tranquillity (III.14.1) and of concentrating all the senses within oneself (VIII.15.1).

Śaṁkara in the *Brahma-sūtra-bhāṣya* did not discuss meditation in many places, but did acknowledge its importance. *Yoga* is the means of attaining the realization of Brahman (I.4.1): the activity of the sense-organs needs to be restrained, and we need to abide within the mind alone. Thus the highest self is presented to the minds of yogins (III.2.24). The seeker of knowledge must possess such calmness of mind, obtainable when the senses are subdued (III.4.27). All this is so because Brahman is not an object of the senses (e.g., I.1.2). The *Viveka-cūḍāmaṇi* states that reducing the world until it seems like dream-images (i.e., a nature-mystical experience) is not enough; the self is attained only through concentration free of all distinctions *(nirvikalpa-samādhi),* where all sense of separateness is ended and the world of appearances vanishes.[32]

Śaṁkara did not stress the need for meditative cultivation to the degree the Buddha did. Concentration is an *action,* while knowledge is not (BSB I.1.4).[33] So also there may be a variety of different experiences, but enlightenment is one and final, admitting of no degree (BSB III.4.52). But to contrast the Advaita and Theravāda traditions on this point—enlightenment through knowledge versus enlightenment through experience—

would be an error. Both maintain (1) the need for a concentrated mind, (2) the difference between mental preparation and the enlightening knowledge, and (3) that knowledge involves transformation, not mere understanding.

Because Brahman is not an object, it cannot be known as an object is (i.e., dualistically). Instead, to "see" the seer of seeing, we need a complete "unknowing" of dualistic awareness (BU IV.4.15): to realize that we are dreaming, we need to wake up—nothing done by the character within the dream will work. That is, no amount of our action in the state of nescience will produce knowledge. A direct realization of nonduality alone will do. The knowledge that is enlightenment is the depth-mystical experience and is itself reality, since the awareness constituting the experience is the awareness also constituting the reality (Brahman). It is not an experience of *union* with Brahman—we already *are* that reality—but is the realization that Brahman is the reality of the world. Release is not something "brought about" or "accomplished"—it is eternally Brahman (BSB III.4.52). Brahman is not a thing "gone to" or "attained" (BSB IV.3.14) or "possessed" by a person. Knowledge "produces" nothing; an epistemological change is all that occurs. The cycle of rebirth is ended, but this is merely a phenomenal change; the reality involved in rebirth remains the same. Knowledge itself, since it is reality, does not change but is eternal. Nothing is accomplished; nothing changes.

With the realization of the truth, all the practical distinctions of ordinary life—knowers, acts of knowledge, objects known, authorities—are seen to be products of nescience (BSB, introduction, I.1.4). All the notions of knowledge and nescience, Brahman and the realm of nescience, are distinctions from within the realm of nescience. So also the distinctions of "release" and "bondage" are part of the view of the world that ends with the replacement of nescience (I.1.4, I.2.6). Nor is there any individual seeker of liberation—there is no individual trapped or free (BS II.32). Liberation consists merely in realizing what has always been the case.

Through desuperimposition (*apavāda,* when an idea attached to something is recognized as false and driven out by the true idea), the nondual ground is seen (BSB III.3.9). It is similar to realizing that the morning star and evening star are identical—and that neither is actually a star. Nothing changes except our perspective. In the sphere of true knowledge the phenomenal world vanishes (BSB I.2.20), but one liberated in life *(jīvanmukta)* will have "wrong knowledge" *(ajñāna)* replace knowledge *(jñāna)* after enlightenment, that is, in the enlightened state one has returned to the dualistic realm of sense-experience with its indefinable status. The senses have returned but seeing diversity, a plurality of real entities, is ended. The content of our waking state is never negated (BSB II.2.29), but now we see correctly. One is perceiving the world of duality (i.e., sensing the stimuli) in the waking state but not perceiving it (i.e., not taking these experiential distinctions as indicative of reality) (*Upadeśasāhasrī* X.13). The "dream" is still occurring, but having awakened through the experience likened to deep sleep, we know that the "dreamer" alone is real and that the content of the "dream" corresponds to nothing real. Questions of the content (how much water is there in that mirage? where did it come from? how fast is it evaporating?) are not answered but rendered meaningless for one who has transcended the situation through knowledge.

PART II

The Nature of Scientific and Mystical Claims Compared

3
Ways of Life and Knowledge-Claims

Part II will consist of comparisons and contrasts concerning the nature of scientific and mystical claims. Only the most central or illuminating aspects of these claims will be covered, the purpose being to reveal similarities and differences in these two enterprises.

Basic Aims of Science and Mysticism

Fundamental to the whole frameworks of scientific and mystical knowledge are these enterprises' respective basic aims or intentions. In this regard the enterprises diverge in a simple and clear-cut manner. Mystical knowledge forms an integral part of total ways of life. The natural sciences differ in their concern: only understanding the physiochemical or biological dimensions of reality is of interest. The differences are reflected in the questions asked and what is acceptable as an answer. The sciences do deal with beliefs about the world and have a legitimate function in informing our view of reality (and consequently how we act). But this cannot become in itself a belief-component of a "scientific" way of life: science can never supply in a straightforward, deductive manner the values and codes of conduct constituting a full way of life.[1] To expand the articles of faith implicit in scientific research or the more metaphysical principles contained in one particular theory into a total conceptual scheme upon which alone actions can be decided is to enter the realm of speculative metaphysics, if it is possible at all.

The problem inherent in connecting scientific theories to life-orientation can be seen by considering contemporary cosmology. Some scientists believe that the "more the universe seems comprehensible, the more it also seems pointless."[2] We might ask, however, what could be found by science that in principle would show that the universe has a point? If nothing, then science could not be the ground for concluding whether the universe is purposeful or not. In fact, the different interpretations given various cosmological positions defended today reveal that science itself does not supply answers to the question of meaningfulness. Three possible ends to the universe are theorized: (1) the components of the universe will expand indefinitely, precipitating a "heat-death" of all life; (2) a collapse of the components of the universe into a primal ball; (3) a reexpansion following the collapse. Many scientists have found life meaningful within an expanding universe. For example, Arthur Eddington would prefer a universe that achieves its purpose through evolution and lapses back into changelessness to a universe that makes its purpose banal through continual repetition.[3] Others see the end of mankind as precluding the possibility of any ultimate meaningfulness to any existence; purposefulness requires that the universe be inhabitable forever. Our meaning is derived only as a contribution to humanity as a whole, not individually. The second alternative does not foreclose the possibility of human significance, if Eddington's position can be maintained; but it does not guarantee it either. The last possibility would permit indefinite inhabitability of the universe (and thus permit human meaningfulness under certain interpretations); but it does rule out any human progress if each oscillation is independent of all others. With the reexpansions we may have merely a pointless series of universes, each repeating the same type of events over and over. Even that our universe produced life could not be cited as evidence that reality has a purpose, since it is possible that in some oscillations there would be no life at all. Buddhists and orthodox Indians have been able to find life meaningful with particular conceptions of an oscillating universe, even if the meaning involves getting out of this realm.

Thus cosmological theory alone does not dictate whether the world is meaningful or not. Cosmology may change drastically in the near future, but the issue of whether our life is meaningful will remain. The issue of the meaningfulness of life depends upon extrascientific considerations. The values involved in any answer will come from other sources, such as the Bible. The role of fundamental values can produce an emotional reaction even among "disinterested" scientists. A pulsating cycle of universes is preferred by many, perhaps because so much else in nature runs in cycles. Robert Jastrow mentions that some scientists are "curiously upset" with the idea that the universe had a beginning because it conflicts with their "articles of faith."[4] Scientists "would like to reject it" and believe "it cannot really be true."[5] The idea of a "singularity" that cannot be subsumed under a law disturbs scientists. And certainly an event behind which they cannot in principle explore—the theory being that the heat of creation destroyed any evidence of what might have come before—also would disturb them. Thus scientists go on searching for the mass necessary for an end to expansion to occur (thus permitting the possibility of a repetition of universes).

Scientific work itself may also provide momentary joy or despair, or even become the source of meaning for many socientists. But a scientist need not be any less effective as a scientist if God or some other religious principle were the ground of his or her life-orientation. Scientists may become better scientists, or worse ones, if science is the central concern in their lives; but such concern is not necessitated by the scientific work itself.

Religious beliefs may be a source of inspiration for scientists. Historically, the relation of science and religion has varied. Religious faith and the biblical doctrine of the world as God's creation may have been factors causing the rise of modern science. Certainly many of the prominent scientists of the sixteenth century saw science as a religious activity—in Boyle's words, "the disclosure of the admirable workmanship which God displayed in the universe." So too Einstein speaks of religious wonder and awe as the strongest and noblest mainspring of scientific research.[6] But religious faith can lead people away from science, too: the lack of science's significance for the matter of salvation may have been responsible for the relatively slight interest in natural science in the Western Middle Ages. Such varying responses should show that there is no religious dimension to science *per se,* but only varying extrascientific reactions. Pierre Teilhard de Chardin may have thought his work was a religious science. But while religious interests may direct the choice of topics to be researched scientifically, still seeing religious significance to the work comes only from an extrascientific framework.

Science's *ethos* is usually depicted as involving hope, humility, honesty, an open mind, and cooperation.[7] Of course, not all scientists are as humble as Einstein—many are as vain and ambitious as Newton. So too fraud and theft have been common throughout history as scientists compete for money, power, and prestige. Nor is science so open-minded even today if Lakatos is correct in asserting that "present mathematical and scientific education is a hotbed of authoritarianism."[8] Political influence can also be brought to bear upon scientists, as illustrated in German opposition to "Jewish science" in the 1920s and 1930s, and Lysenko's control of Soviet biology later.

The ideal *ethos* of science could be generalized to govern one's way of life. Certainly the optimism of science is mirrored in the greater cultural faith that the progress of science and technology will transform the world for the better. Nevertheless, these values cannot supply all the criteria upon which to make choices. The connection of science to the military, a connection that has always been present but that has increased greatly today, regardless of the political system in force, reveals that science does not select all the values necessary for a total way of life. Accepting science as providing a cognitive standard is a value-judgment all scientists make. So too value-choices of a moral kind are required: choosing which topics to study—a problem especially great in biology and any science affecting public safety—or imposing any other limitations upon research arises from considerations other than the inherent value of the topics from a strictly scientific point of view. Scientific topics and interests do not dictate what *must* be studied at any expense and regardless of consequences. It is the scientists (whether as scientists or as human beings) who may be moral or immoral.[9] It is a legitimate question whether scientists should be held accountable for consequences of applications of their research. Any "mad scientist" who commits an act "because science demands it" is operating with an implicit value-commitment not required by science itself. Selecting science as a value in making decisions or promoting its development in a society is itself a value-judgment and must be justified as such.

Since science does not contain all the values necessary for a way of life, it cannot become a way of life. Even extending any values from the scientific *ethos* to all of one's life involves a value not supplied by science. So too mystical enlightenment changes one's way of life (or at least one's dispositions), while a switch in theory may not affect the scientist at all. That science, unlike mystical ways of life, cannot answer the question of how to live does not preclude the possibility of significant commonality with mysticism in other regards. But the differences in aims does preclude science from ever becoming mystical in *aim* and *nature:* science does not encompass the experience or dimensions of significance that mystical ways of life do. This is as true today with the fundamental changes in physics as it has been in the past.

In principle, science can be incorporated into a mystical way of life: scientific theories may be accepted or scientific research may be engaged in as an exercise in mindfulness similar to gardening.[10] Probably any natural object or idea can touch off a numinous or even a mystical experience. Simple contemplation of the stars and of astronomical facts is a commonly cited source. Nevertheless, this does not make astronomy or any other science a systematic attempt to achieve mystical enlightenment. Pictures of cloud-chamber interactions are aesthetically attractive to many people. But this does not mean that their intended scientific purpose or use is as art. They can be so treated, but their scientific function is entirely different. So too a scientist may practice meditation and study mystical belief-systems in his or her spare time; meditation may help one's energy or concentration and make one more disinterested in possible results, and the mystical systems may suggest new ideas workable into scientific hypotheses. But this does not make mysticism a natural science. Nor does this make science mystical; the concerns still differ, and science still values the increased working of the analytical powers of the mind. In such circumstances, science cannot be looked upon as a spiritual way of life or a way to enlightenment, nor can scientists be considered spiritual masters.

The Relation of Knowledge-Claims to Cultural Phenomena

Mysticism and science share some features as cultural phenomena. To begin with, they are both *social* in nature: there is no completely individual mystical or scientific system because every person is born into a society containing a matrix of learned and handed-down belief-claims and practices that confront anyone attempting to understand new experiences; and only the socially accepted understanding survives.

In addition, both enterprises arise out of a fundamental human need for order. To avoid danger and provide for security, reliable patterns in the flux of events are emphasized by different cultural creations. Cultures as a whole transform physical reality into our experienced reality; part of this makes them into vehicles for coping with the complexity and insecurity constituting part of suffering that Buddhists discuss. Peter Berger speaks of the nomizing urge;[11] Clyde Kluckhohn characterizes myths and rituals as supplying fixed points in a world of constant change and disappointment;[12] Mary Douglas similarly discusses a need for order, especially to avoid chaos at the margins of life.[13] The primacy of a basic orientation to cope with the contingencies of life has been asserted by some psychologists.[14] Mystical ways of life provide the orientation and meaning necessary for survival. The sciences, on the other hand, cannot do the same; still, they provide order of a more limited scope related to survival, material needs, and understanding our place in the world.

The parallels Robin Horton finds between traditional African religious thought and Western science (introducing simplicity, unity, order, and regularity into what appears complex, diverse, disordered, anomalous, and so forth)[15] reflect merely this universal human need for ordering. It is sometimes claimed that scientific knowledge arose out of religion or magic. But considering the differences in ranges of experiences covered and in the types of questions asked, it seems more accurate to say that both science and religion have arisen out of a common need for order. There is no reason to believe that science would not have arisen without religion.

What is considered satisfactory understanding or knowledge depends upon the *Weltanschauungen* of cultures (i.e., their general conceptions of the "scheme of things") upon which the people are prepared to shape their actions. If one ignores the issue of whether there are really any truly integrated cultural "forms of life," ideally a person's vision of the fundamental nature of reality provides the broadest frame of reference or patterns of intelligibility for use within all more limited conceptual schemes. It underlies specific factual claims in the manner that equating the death of a person with the cessation of all brain activity reveals much about what we take a *person* to be. World-views are revealed more by the type of analysis involved (what questions are asked and what counts as an answer) than by particular claims.[16] Their most concrete expression is usually given in the religious system of a culture through images, stories, examples and so forth.

Each world-view, each conceptual "world," has latitude in determining what is to be enshrined as real. What exists independent of us is one component in any construction of such conceptual world. If we work with the tacit assumption that the earth is flat and has edges, we would still not be able to find any edges to sail off. Even if social constructions are not absolutely arbitrary and an infinite number of configurations are not possible, nevertheless choices remain. We can survive in the world with a great variation in world-views and values, as the variety of cultures throughout the world indicates. Mystical systems, even within one culture, show the same variability and limitation of alternatives of conceptual and valuational schemes. Conceptual elements (nonduality, reality) and values (e.g., nonattachment) are shared, but the basic concepts and claims, when incorporated into whole systems, vary.

Science too may be in the same position. As mentioned in the first chapter, science is affected by the broader world-view of the culture within which it is set. Sociologists of knowledge will concede that social and cultural factors external to the history of science itself affect the development of each science. But for sociologists this influence is limited to the amount of impetus for the development of science and to the inspiration of scientific ideas: the final acceptance of theories, that is, scientific knowledge, depends exclusively on arguments independent of cultural influences. In this way sociologists of knowledge exempt natural science from any possibility of cultural relativism. Thus the Pythagorean influence upon Copernicus is irrelevant to the truth of his astronomical theories. Certainly not all theories and world views are mere mechanical reflections of social interests.[17] Evolution cannot be dismissed because Darwin was influenced by Malthus and the socioeconomic structure of nineteenth century England. All scientific intuitions presenting new ways of ordering phenomena are to a degree free of culture. Nevertheless, cultural factors do figure in the final determination of which factors are significant—even to determining what the "facts" are—and which explanations are acceptable or unacceptable.

Science in turn interacts with a world-view: the world-view provides the ultimate context of justification and science reacts back upon it as it develops in the light of new experiences and ideas. Science, like religious belief, becomes part of the common sense as it shapes how we see the world. The world-view exerts a conservative pull upon science because the ultimate accountability of a theory involves the total cultural knowledge of the time. But science is able to effect revolutions to the extent that it seems odd even call the idea of the earth's moving a *theory*. Such is the change in common sense.

Religious belief is one possible source of influence on science in two ways: it may be a source of ideas, and, since religion is a major cultural factor, it may affect the grounds for accepting a theory. As scientific conceptual schemes come to encompass more of our experiments related to subatomic and cosmological realms, their breadth will make them more susceptible to influence by religious cosmological ideas. Perhaps this is the reason Huston Smith speculates that the science of the future will be more like religion than our present technology.[18] As our own culture evolves, religious traditions from Asia may have a major impact in this area—on what we accept as real—and therefore affect the direction of science. (Again this does not mean that science is becoming more mystical in *nature*).

Science's strong impact on the evolution of Western culture until the present is a matter of historical contingencies, not of logical necessity. Its history is not of uninterrupted, linear progress. Science has attained its importance and degree of development in the West because of particular cultural values, not some innate superiority it holds over other human enterprises. The roles of the ancient Greeks' predilection for seeking order in events and of the biblical conception of the world as the ordered creation of God are often cited as its historical roots.[19]

It could be that science has a universal character, that is, that when the proper social conditions arise within any culture uninfluenced by other cultures already having science, science will always have the same aims and eventually develop along the same lines as our "Western" science. But the possibility of the cultural relativism of science must be conceded: the theoretical component and therefore the conceptualization of experience may be entirely distinct in different cultures and always remain so, even if there is a transcultural "scientific method" independent of all theories. All societies have some empirical knowledge of regularities in nature—otherwise they would not survive.[20] In archaic cultures, most explanations are present as a dimension of the religious system, not as a separate scientific enterprise (scientific explanations and theories arising only when cultural phenomena became more compartmentalized). The refinement of this empirical knowledge will be along lines established by the explanatory aspects of the world view: what direction research takes is determined by what one takes to be the significant problems or questions, and how the data will be ordered depends on "ideals of natural order" and even more basic metaphysical beliefs. More important, the standards for the acceptance of a theory are not timeless and nonculturally bound. The explanatory component cannot be separated in a hard and fast way from the broader conceptual framework of a culture. Fitting in with the basic visions of reality is one desideratum of theory-acceptance, and the application of the others will also depend upon the cultural values.

The traditional sciences of India were heavily influenced by Middle Eastern sources and were not closely integrated with religion. How Indian or Chinese science would have developed independently of Western influence and given the proper "scientific" interest (an interest in physical and biological regularities independent of other dimensions of significance) is an open issue. It may be that all science would eventually reach the state of our "Western" science. Or, because of different visions of reality guiding theory-construction in different cultures, there may very well have developed a cultural pluralism of rival theoretical frameworks, each of equal insight and value. Only the lowest level of arithmetic, if even that, may be truly transcultural. In any case, Western culture has polluted any "pure" non-Western scientific traditions. The actual historical situation is rendered more complex by the fact that there have been cultural and specifically philosophical contacts between cultures since before Alexander's conquests; for example, Pyrrho of Elis may have obtained his version of skepticism from Jaina monks in India. But it is difficult to identify any specific South or East Asian influence or elements in our natural science (although the "Arabic" number system comes from India and the zero probably developed in Northwest India or China). Today Asians are making significant contributions to mathematics, cosmology, and other sciences. But these scientists are trained in "Western" science.

Detecting specifically *mystical* elements of influence on science is even more difficult. Science in the Renaissance did evolve through the impact of "occult" movements, although not mysticism as defined here.[21] Renaissance Platonists' interest in the eternal changelessness of mathematics, Kepler's interest in astrology, and Newton's interest in alchemy are instances of this impact. Today one instance of the possible influence of mysticism upon physics is Schrödinger's study of Advaita Vedānta before he became a physicist; this could conceivably have directed his scientific thought, although no specific points of such influence have been detected. Other seemingly mystical ideas in science (e.g., Einstein's emphasis upon simplicity and unification) may be better understood as cultural factors common to mysticism and other enterprises rather than as derived from mysticism itself.

4
Reality

Speculative Metaphysics

One broader conceptual system of special interest is metaphysics. "Metaphysics" has several senses, which need to be distinguished. Speculative metaphysical systems are all-embracing conceptual systems resulting from self-aware reflection upon the various types of experience and claims. Each delineates the most general structures of reality by placing the insights from each area of human experience deemed important into a systematic, ordered whole. Thus each deals with the "ultimate nature" of reality, that is, our final verdict about the nature of reality once all considerations that we accept as important are taken into account.

Metaphysical speculation involves giving one type of insight central importance and developing a vision of reality on a scale grander than that of any one field of study. "Root metaphors" are utilized to comprehend all the complexity of reality in terms of some familiar experience, event, or entity that is considered paradigmatic of the fundamental nature of reality.[1] Two metaphysics popular in the modern world have taken their metaphors from technology and science: from Newtonian mechanics, reality construed as a machine (isolatable parts operating only externally upon each other in an absolutely law-governed manner); and from biology, reality as an organism (the world as composed not of distinct entities but of interacting parts of an organism). None of these positions, nor any other form of speculative metaphysics, is dictated by any scientific theory, but all are the results of other considerations and judgments.

The criteria of acceptability used to construct these systems should be those of scientific theories: simplicity, coherence, and so forth. The problem of how to assess these systems is rendered more difficult because the criteria apply to theories covering all of reality, not theories within one limited domain of study. More emphasis is placed upon comprehensiveness and completeness in covering each domain than is the case with scientific theories. So also metaphysics may go radically against our commonsensical views in order to give a more insightful and accurate account of the way things really are (science of course does this too). Deciding between metaphysical systems is more difficult than with scientific theories because finding agreement on how to apply the criteria of acceptability between competitors is even more difficult. Experiential falsification has much less strength here than with science: metaphysical principles are of such a general order as to be reconfirmed no matter what occurs and thus are most resistant to disconfirmation. Hence, such systems seem to be self-confirming: if the system gives an accurate account of the ultimate structure of reality, then all events should be as they are. Predictions of phenomena to test one metaphysical system over another are not possible. Metaphysical systems do fall, however, with the rejection of *theories*. That is, systems die when, in the light of accumulated new experiences and specialized explanations, the root metaphor no longer appears insightful.

How are mysticism and science related to speculative metaphysics? First the mystical systems. Religions are concerned with the status of more than one type of claim (moral, historical, experiential), but are not necessarily tied to thoroughgoing metaphysical speculation. Both religion and metaphysics place more than scientific understanding in a central position, and thus there is a natural tendency for religions to adopt some metaphysics—and a tendency to accept metaphysics as essential to the religious tradition. For example, Christianity has been connected to Platonist, Aristotelian, and Whiteheadian thought. Still religions need not attempt a systematic thought-construction, placing all areas of experiences into one whole. That is, the world-view as a component of any way of life need not answer every metaphysical question that might come to mind in order to be acceptable. Metaphysics is simply a more intellectual, speculative enterprise than religious thought needs to be. Even if one ignores the evolution of thought in a religious tradition, religious texts are not even entirely consistent (this is especially true of Theravāda Buddhism).

But mystical systems are instances of religious systems that do tend to be speculative. This may result from the mystics' unusual experiences: placing as cognitively central an experience that contrasts greatly with everyday and scientific experiences may give rise to the need to establish a comprehensive system fitting all experiences into one total system. At a minimum, in denying so much of what we normally take as real and certain, mystics become more aware of the problems involved in belief.[2]

Whatever the reason, mysticism has been more speculative and rationalistic than science. Our need to find patterns leads to a tendency to value the changeless underlying the flux of change. Identities of seemingly discongruous phenomena are present in science and even more so in metaphysics. This is the source of "entities," "selves," "substance," constructs we use to simplify our account of reality. In Vedic India the older *Upaniṣads* contain multitudes of identifications between the powers behind diverse types of entities (e.g., the earth, speech, and the stars with the *Ṛg-veda,* or the sun with the eye). Eventually, the inner self became identified with the "self" (substance or essence) of everything in the universe; thereby a changeless ground was established. The Advaitic claim that the self constitutes the object of sight (BSB I.3.13), and its intricate arguments revolving around such root metaphors as the magician and his trick to counteract seemingly negative sensory evidence, are parts of what is probably the paradigmatic instance of a metaphysical system.

The discussions of the Buddha in the Pāli canon are not so systematic in their treatment of issues as Śaṁkara's commentaries on more organized treatises. The Buddha also dismissed certain issues as irrelevant to the religious problem of suffering and as based upon reasoning ungrounded in experience. Nevertheless, the Theravādins do make claims of a speculative-metaphysical nature even if they have no complete system. Most important are their ontological claims: the Theravāda metaphysics is based not upon "being" but on "becoming," that is, on change and a denial of permanence. The root metaphor for this tradition is the chariot-analysis (S I.135). The same experiences (mystical and otherwise) as interpreted by the Advaitins are interpreted by the Theravādins in a different manner. For example, each tradition goes against our Cartesian common sense—Advaitins equate any self-awareness with Brahman, and the Theravādins view it as episodic with no mental substance behind it.[3] Each has the same base of experience and each fits its view into a larger account of reality. The Buddhist metaphysical framework substitutes a flow of dependently arising factors in the place of unchanging substance. The accompanying problem is how to account for regularities within the flow of phenomena (e.g., separate lines of continuity instead of "persons") without appeal to something other than these factors.

In the first chapter it was mentioned that science is not itself a form of speculative metaphysics. Taking science as the only source of truth and the final depiction of the entire structure of reality goes beyond the sciences themselves into metaphysics. Still, science under the realist interpretation does involve claims about the nature of the world. But in maintaining that some actual features of reality are revealed in laws and theories, no claim to have discovered the "essential nature" of reality or the sum of a phenomenon's reality is implicit. If scientific analysis reveals something of reality that is independent of and irreducible to other features, these findings have a claim to ultimate reality. But scientific analyses can be reconciled with different root metaphors. Just as the subject-object stance implicit in science can be reconciled to various psychological monisms and dualisms concerning the relationship of "mind" to "body," so also scientific results as a whole can be embedded in various metaphysical systems. The broadest articles of faith in science can be incorporated into the principal speculative alternatives. No commitment to a real plurality of entities or any other stumbling block is required by the nature of the analysis. Only solipsism is incompatible with scientific realism.

Prescriptive Metaphysics

Another sense of "metaphysics" is also relevant to the matter at hand. It is that metaphysical principles, even on a scale less than a speculative synthesis, frame how experiential phenomena should be seen. They are prescriptive rather than simply descriptive in the sense that only certain aspects of phenomena are highlighted as real or at least as most important to a particular concern. A description, on the other hand, would be an account acceptable to all regardless of one's broader considerations—something that might be true only on the levels of understanding most directly tied to sensation.

Saying that feature *x* of the phenomenon is central to a scientific or a mystical concern is not to claim that *x* is its complete essence or ultimate nature. Thus systems of thought need not be speculative metaphysics to have the property of prescribing certain visions of reality or points of view from which phenomena should be viewed, although speculative metaphysics may have this feature too. But while speculative metaphysics involves more how we understand things, prescriptive principles are more directly related to how we should *perceive* reality. Speculative metaphysics is comparable to the situation of a stick looking bent in water: only our knowledge is affected, not our sense-experience, when we correctly understand the situation. Prescriptions are more like suggestions on how we ought to view Gestalt figures. This type of metaphysics can become speculative. That is, from the prescribed point of view, we see phenomena "as they really are"; other aspects are at best inconsequential and at worst subjective distortions.

Science as a whole is prescriptive: it assumes that we can understand the world without consideration of moral, religious, or other dimensions of significance—that phenomena shorn of all but physical or biological significance reveal something of reality. This determines what is accepted as an observation, what is a cause, and what is accepted as real (a "fact"). Furthermore, a stance is prescribed: a subject-object dualism. Perhaps additionally a state of consciousness is also prescribed: the reproducible conditions required for scientific observation may need a "normal" (dualistic) state of consciousness. Within specific theories too, the metaphysical elements act as policies or methods regulating theory-construction.[4] In the Middle Ages, essences and God's purposes, not mass and force connected by the laws in space-time, were the primary concern. Early modern science made particles, force, space, and time the primary categories. Today the concept of "fields" has rendered "action at a distance," and even "force" in certain contexts, obsolete.

Religions in general and mystical systems in particular are regulative in this sense too. (In addition, they prescribe goals and appropriate courses of action). Religious believers more easily slip into treating their beliefs as absolute—that they see reality as it really is to the exclusion of other points of view—but this is not necessary. The systems differ in what aspects of phenomena they value. For the Advaitins and Theravādins, the favored elements are easily discerned. Śaṁkara favors what does not change: permanence is central to his conceptual scheme. He also attempts to provide some understanding of the mystical. The Buddha, in emphasizing impermanence and change in the flux of experience, would leave any questions concerning any alleged mystical reality unanswered. Their judgments concerning phenomena differ correspondingly. In J. F. Staal's words, we can attach higher value to the material cause (e.g., clay) and look upon the form (e.g., a jar) as ephemeral, or we can evaluate the form more highly and speak of the creation of something new; the evaluation is dependent upon us, not on the phenomena.[5] The weak point in each conceptual scheme is the inverse of the strong point: in Śaṁkara's, why is there any appearance of change at all? And in both Śaṁkara's and the Buddha's, why are there regularities throughout change? Why can we formulate natural laws that are so successful in predictions?

For a more detailed example of prescriptive metaphysics, consider Buddhism. That the Buddha's doctrine is prescriptive in nature has been suggested by others.[6] Instead of giving neutral descriptions which everyone would agree to, the Buddha indicated a specific way of viewing reality, a way leading to the end of craving. The Buddha is a religious "revisionary metaphysician"[7] evoking a reorganized apprehension of reality (although he is not giving a complete speculative system beyond his concern with suffering).

The Buddha's teachings entail a specific vision of the scheme of things. That is, looking at the world with Buddhist "right views" is not looking at it with *no* views. When one looks at the world in a certain way, many sets of questions and associated actions are rendered purposeless or become actual hindrances in the religious quest. But it does not follow that we are not looking at the world through a particular perspective. Furthermore, introducing the idea of "as things really are" introduces speculative metaphysics. Any view advanced *outside* the designated enlightened framework will be branded wrong; it appears to be an unintentional or intellectual misinterpretation of what is truly there. On the other hand, once the prescribed experience has occurred (i.e., once a person is "inside" the prescribed way of looking), the point of view no longer appears to be a *view* at all but rather the way things really are.[8] To the insider (the enlightened), it always is the outsider who sees the rope as a snake, while we insiders see only what is there. We are tough-minded, they subjective; we see what cannot be denied, they distort. It is not so much audacity as simply part of the logic of speculative metaphysical belief that this is so. We naturally take our view as seeing reality correctly rather than as seeing reality within a conceptual scheme. Buddhists do not think they are seeing the stream of experience *as* completely essenceless but believe they are seeing *that* the stream is that way.

All the Buddha's teachings are geared toward giving a perspective through which we can see our experiences with the proper insight, so as not to arouse craving and thereby cause suffering and rebirth. The no-self doctrine, dependent origination, and so forth set up pictures of reality that act as viewpoints for seeing all experience and the world in a new light. Saying "Look upon the world as a bubble; look upon it as a mirage" (Dhp 170) or "Consider the whole world as void" (*Sutta-nipāṭa* 1119) ought to be seen as just that—suggestions, a form of *persuasion* (albeit one entailing specific knowledge-claims).

To see the interaction of the conceptual scheme and experience, consider perception in the state of mindfulness. This perception is praised as seeing reality "as it really is" *(yathābhūtaṃ).* It is allegedly a state of sensory experience with all cultural encrustations removed, so that we see the flux of reality in a pure state, the only restrictions being those we have as human beings with a particular sensory apparatus. But we can see that structuring elements are in fact present in this sense-experience by examining Mahātissa's enlightenment. This disciple of the Buddha became a "worthy" *(arahant)* by realizing the impurity of the body while gazing at sunlight flashing off the teeth of a laughing woman. Later, in answer to an inquiry as to whether a woman had passed by, he replied:

> Was it a woman or a man that passed this way? I cannot tell. But this I know: a bag of bones is traveling on this road.

Buddhaghosa in the *Visuddhimagga* states that Mahātissa took no note of signs and tokens. Signs *(nimittā)* are explained to be anything arousing the passions, and tokens to be personal characteristics that indicate and reveal the passions.[9] Why does he not see what might arouse the passions? Not because they are not there. Following the chariot-analysis, "men" and "women" may not ultimately exist, but their parts, the *dhammā,* do, including those parts which arouse passions. They are as real and effective as any other parts. In the Abhidhamma, the faculties *(indriyā)* of feminity *(itthi)* and masculinity *(purusa)* are as real as the eye, nose, and so on. Furthermore, Mahātissa saw a "bag of bones"; his was a differentiated perception, not totally unstructured visual chaos. (There were aural perceptions as well, since he was able to understand the question asked him and was able to answer.) Therefore some element of structuring must have been active. More is involved in this "right understanding" than seeing all factors comprising experience as void of substance: specific factors are being excluded rather than all factors being perceived in a new way. Mahātissa has internalized a screen barring any factor that might give rise to an emotional response. Indeed, the mindfulness-exercises decide which ideas and even which perceptions will be allowed into our awareness.[10] It is not impartially "neutral."

Thus there appears to be no reason to doubt that the state of mindfulness involves experiential and conceptual elements like any other observation.[11] This intentional awareness may involve a Gestalt-like switch in which a new configuration of "facts" is formed—the sensory stimuli remain the same, but they are structured in a new manner by elements from the mystic's *internalized* conceptual framework (as suggested in chapter 2). During the earlier destructuring stage there may be a period in which no structure at all is applied—a bare mindfulness free of all conceptual control. But in the *enlightened* state (the state after the enlightenment-experience in which a permanent character-change has occurred), some structure is always present—we cannot live for long in a world of pure sensory flux; some guiding abstractions are necessary. But there does not appear to be exactly one enlightened conceptual system or course of action; the differences reflect concerns and interests not directly connected to the mystical experiences. For the Theravādins, the emphasis is upon impermanence. The analysis sets up a general screen upon reality that emphasizes arising and falling, and the division of apparent unities into parts. But this involves making lists of pigeonholes and a naming process that appears from other points of view as arbitrary as any other. Why should one step or any organic process be divided into exactly three phases, for example? The elaborate lists of the Abhidhammists are a natural outcome of this approach.

Once we have totally internalized the doctrine, the "truth," that is, knowing things as they really are (D I.83), of course becomes what is denoted as basic in the conceptual framework. Our awareness reflects the prescribed analysis and thereby seems only to confirm that this is reality as it is. "Right views" may be impartial in some regards, but they are no more complete than any other view. But for Buddhists, dependent origination is not merely one way of looking that is oriented toward specific concerns, but reality pure and simple; and the *dhamma*-analysis is not a prescription but a description that should be agreed upon regardless of one's theoretical point of view. With such certitude, situations can be dealt with spontaneously without the mediacy of reflection. This absence of argument does not mean that one sees reality free of conceptualization, but only that the role of new concepts in the process is not apparent.

The normative character of the Buddha's religious program should be reiterated. There is a mixture of pleasure and pain throughout our lifetime, but the Budda emphasized only one (suffering) to the exclusion of the other for soteriological reasons. K. N. Jayatilleke can make the claim that the Buddha is being realistic in advocating *nibbāna* "even in the face of the fact that as the Buddha says 'human beings enjoy on the whole more pleasant experiences than unpleasant ones' "[12] only because of this metaphysical reasoning. Why should we not accept the impermanence of all experiences and cultivate mundane pleasant experiences? Obtaining a state totally free from suffering may be a worthy goal, but emphasizing the painful over the pleasurable is not a neutral appraisal. The soteriological value in treating the body extremely negatively in meditation—viewing it as a boil (S III.168), the corpse-meditations, and so forth—is obvious: it reflects the desire to be free of it.

Such forms of mindfulness involve a shaping of the mind. The training to feel pain and pleasure without permitting any emotional responses ("While I am walking up and down, covetousness and grief, and other evil, unsalutary thoughts shall not enter my mind")[13] requires great effort. Similarly, seeing reality without reference to a self or objects also is extremely hard. Just because special practices are involved, it does not follow that such perception is not an insight into reality, but it must be argued that such effort does not induce a delusion. Contemplating the impermanent to abandon any notion of permanence and so on is the same type of exercise as seeing all objects as blue or red (D II.110) or experiencing an unpleasant object as pleasant or neutral (S V.119). Are any of these "things as they really are"? This might be behind Śaṁkara's attack on any meditative identification as an instance of superimposition.

In some contemporary meditative experiments, perceiving involves a loss of a third dimension.[14] Is this how objects actually are? Observing a stick in water, it appears bent; relying upon such observations alone, should we conclude that the stick is really that way? If during meditation we should abandon concern with the future and the past, and concentrate upon a phenomenon in isolation, does this mean that temporal and causal relations are unreal?[15] Can we speak of the way things really are apart from the processes behind appearances? This string of questions at least raises doubts about the self-evidence of meditative claims. Mystics need to advance reasons for accepting their perceptions as cognitively more fundamental than ordinary ones. And more generally, it might be asked whether any prescriptive account of the world can lay claim to depicting reality as it actually is to the exclusion of other accounts.

Criteria of Reality

Central to any enterprise that sets about to describe how reality truly is is to characterize "reality." "Reality" or "really exists" is employed in everyday usage to contrast with appearance, illusion, or deception: we talk of something as real only in disparaging something else (BSB III.2.22). Thus if we think that we see something but all other information contradicts it, we dismiss the content of that experience as unreal. A minimal definition of "reality" is, following William James, what cannot be denied regardless of our other experiences, what must be taken into account, or what is to a degree beyond our control.

The sciences themselves do not give a full metaphysical characterization of reality, but the realist interpretation does maintain that they are investigations into reality. Within the mystical traditions, discussions of the nature of reality are more common. The prevalent criteria used within mystical systems and philosophies influenced by science can be grouped as follows (accompanied by criticism from the points of view of other criteria):

(1) Fundamental independence: something is real if and only if it does not depend for its existence upon, or is not reducible to, something else; the existence of something else in no way affects its reality. A problem with this criterion is that some phenomena (e.g., living organisms) seem to depend upon other realities as the condition for their arising but still maintain a unique level of reality. Problems might also be raised concerning lawful relationships: is the relationship (structure) itself real, and does it depend in some sense upon the related items?

(2) Permanence: to be real is to be unchangeable. If something is ephemeral, we tend to discount its claim to reality. In mystical systems, this often leads to the concept of one fundamental, timeless, indivisible reality upon which every phenomenon is dependent. Everything else is unreal, although how the changeless can be the source of the ever-changing is a major problem. Within scientific and everyday concerns, this criterion leads to concepts of "substance" and "material objects." Sidney Morgenbesser, following Dewey, speaks of fists as being "less real" than hands because they are temporary stages of them, and of bodies as being "less real" than the microscopic entities of which they are composed for the same reason; the latter are real if they are permanent and everything that is can be viewed as composed of them.[16] Difficulties here arose when scientists found on a submicroscopic level the idea of any material permanence to be unwarranted, and found with relativity theory that mass changes with velocity. The "essence" of matter is not permanent extension or anything else changeless. More generally it may be asked why permanence is a necessary condition for attributing reality even if, as seems reasonable, it would be sufficient. As Ernest Gellner asks, what is so marvelous about weight and substance anyway? Is it just a prejudice of ours?[17]

(3) Invariant patterns: if we accept the criterion of permanence but reject the idea of substance, then the organizational structure of the flux of experience is what is real. All states are impermanent and dependent, but the lawful patterns of phenomena are real.[18] Structures and patterns show more permanence than what "fills" them. The background for this criterion is that science is concerned with such relationships. But there can also be regularities and patterns permitting predictions in dreams. In addition, it may be asked why something outside such patterns should be deemed unreal.

(4) Effectiveness: if we are dissatisfied with the passive nature of substance and patterns, we may accept the real to be what is able to bring about an effect (*Wirklichkeit,* "what works"), as with the interaction between sub-

atomic particles. An objection is that we usually do not consider hallucinations by definition to be real, but they can have an effect on the experiencer. If the criterion is modified to include being affectable, mystics would dissent: the mystical is seen as a "power" that is the source of the phenomenal world; it brings about the latter without being affected by it.

(5) Independence of human perception: to be real something must exist outside human experience, whether permanent or not. Our sense-experience of an object constantly changes, but what is experienced does not change (or changes much more slowly). The problem here is that reality must then transcend the known phenomena. To explain sight, sound, and so forth, reality must be invisible, soundless, and so on. Reality in itself would be unknowable or at best known indirectly only to a limited extent. Criticisms from mystics would be that absolute reality is experienceable in the mystical experience and that the conceptual understanding reflecting our everyday sense-experience alone is erroneous.

(6) Experience as reality: all we are directly acquainted with is sense-experience and other inner events; thus they make up some if not all of reality even if they are transient. This may be given an instrumentalist interpretation: experienced phenomena alone are real; scientific and commonsensical objects are only conceptual links. Or at a minimum, the denial of pain, sensory awareness, and so on as real is an error. For the Theravādins, consciousness is as real as matter: each actually is impermanent, consisting of substanceless factors that ultimately are more like qualities than atomic bits of matter. For some mystics, only the mystical experience is absolutely real; "objective" reality is constituted by this consciousness. The impermanence of experience and the metaphysical belief that consciousness depends upon the brain are the chief drawbacks to this criterion of reality.

(7) Intelligibility: "The real is the intelligible, not the observable."[19] Reality is not appearances or observed regularities such as the retrograde motion of planets or a stick appearing bent in water. Instead these phenomena are clues to the reality "behind" them, and getting to that reality requires interpretation. In the final analysis, our understanding, and hence intelligibility, becomes central in what we accept as real. Theories and concensus then become central in defining what is deemed real. Mystics would question whether reasoning should be so central. More generally, our cultural constructs represent what is convenient to us—and it is easy to doubt our convenience as a necssary criterion of reality. Scientific and other concepts are abstractions for special and limited purposes. Their reliability may be indicative of our touching reality, but disorder and anomalies are not therefore unreal. Other objections to this criterion include whether concepts are indicative of reality at all and the variations in intelligibility-standards in different cultures and metaphysical systems. More generally, why must the ultimately real be at all intelligible to us? Does this reveal only our own narrow-mindedness? Or could we ever accept that?

Being and Becoming

Each criterion has strengths and weaknesses. Conflicts are illustrated by the differences between those traditions emphasizing change and effectiveness and those which treat permanent "being" as fundamental. "Being" in mystical contexts means "reality," reality outside of the phenomenal realm of time, change, and differentiated entities; its only "power" is as the ontological basis of everything. Within Western philosophy and more mundane discussions, "being" usually means an abstract universal (what everything that "is" has in common) or merely the instantiation of concepts.

Both science and mysticism distinguish appearance and reality. Appearances deceive—the moon looking bigger than the sun or stars—and reality is to be found beneath them. Nature-mysticism and science each emphasize "becoming" (change) over being: the former emphasizes unreality of any permanent, independent entities; the latter emphasizes the regularities between changes. Neither accepts appearances (products of awareness) as the ultimately real: something more is behind them.[20] Pierre Duhem, in a mystical-sounding phrase, speaks of science as stripping reality of the appearances covering it like a veil in order to see the bare reality itself.[21] But there are differences: the deception for mysticism is accepting a multiplicity of real entities—the "object-ness" is the deception—not the more limited problems of science within the phenomenal realm. So too the scientific reality is unexperienceable. For science, there are genuine differentiations on both the surface and in the scientific depth, and reasoning and ordinary experiences are considered the most valuable means of insight into reality. For depth-mysticism, reality is directly experienceable but not through sensory awareness and is devoid of all properties present in the phenomenal realm. The scientific structures, no matter how unifying, must account for why one state of affairs is the case and not another. The depth-mystical may appear to be a scientific posit of that type: to explain diversity, we need to appeal to what is free of all diversity. But the depth-mystical cannot do this since it is free of all structure and plurality: there is nothing about the changeless mystical that would permit us to expect one particular arrangement of facts rather than another. Any "fourth dimension" beneath the "surface" three would still need structure if it were to be an explanatory posit, and the depth-mystical has no structure. Thus what might appear to be the limiting case of a scientific explanation cannot be such a posit.

There is nothing primitive or unsophisticated in searching for such oneness, although the quest differs in science and in mysticism. Within science, oneness may refer to uniformities of nature (e.g., all electrons are exactly alike) or to the cosmos as a "*uni*verse"—an integrated whole. But it is a process of unification of *structures* that is prominent in science. Within nature-mysticism, it is the unity of *being* that is important—the beingness of reality apart from the conceptual segmentation of reality into entities.

Oneness in this context is the world as a whole, a very big object, as it were. The real for most nature-mysticisms is the interdependent whole, not the constantly changing "parts." The beingness of each dependent entity is the same as that of any other; thus we see *all* of reality at each point in this respect. The Theravādins analyze experienced reality into parts—the wholes, as in the chariot-analysis, being unreal—with little attention to properties *in situ:* they are "atomistic" in that sense, even if the "atoms" (the *dhammā*) are not independent or permanent.

The oneness in depth-mysticism is a complete disjunction from the uniformity and totality of science and nature-mysticism. It is the oneness of absolute *identity*: there are no parts, not even the appearance of diversity due to sense-experience. In Advaita Vedānta, this one reality is not the source of the many, nor are appearances transformations of the one. Instead, the mystical is the only reality. Phenomena do not participate in the real but *are* the mystical nonextended reality. How the one gives rise to surface-appearances is not explained. The *Upaniṣads* more usually hold a view of "emanation" from and "retreat" into Brahman rather than that of the Advaitin identity. The self is one, then twofold, and so forth (CU VII.26.2). The *Śvetāśvatara* (IV.2) relates the one and the many in this manner: "he who is one without color, by the manifold excercise of his power distributes many colors for his hidden purposes." Problems arise here, too: how from its oneness the mystical can become many while remaining itself, that is, remaining undifferentiated and complete through emanation and differentiation. But the problems of how the changeless produces the realm of apparent changes are different for Advaita, where the phenomena are not equally real.

Time

Treatments of the nature of time and space within science and mysticism also illustrate how they compare and differ in their claims. Time involves ordering the succession of occurrences, and space involves physically ordering objects without reference to time. Because they are such unique and fundamental ways of ordering, it is difficult to define either one more exactly without circularities; specifying particular places and times is easy enough in the everyday world, but discussing the nature of space and time is not.

Our sense of time is tied to an awareness of change. Without change we would experience reality as timeless (if it makes sense to speak of any "experience"). Our experience of time is in two basic modes: (1) the succession of events, and (2) duration (the background enduring "flow" of time, the continuum within which change takes place). The latter gives rise to the concept of time as a substance that flows of its own accord, and this in turn produces philosophical puzzles related to time-travel and at what rate that which measures the flow of change itself flows.

We order events in terms of before/after or past/present/future. The former is usually considered more real or objective because it does not depend directly upon our own awareness as the latter series does. Still, before-after gives only a sense of becoming, not of duration. We measure the duration by changes or, more exactly, by repeating patterns (whether natural or social). Only this gives uniformity, since our experience is far from uniform—time "passes slowly" or "flies by" depending upon the events occurring.[22] The nonexperiential method for measuring how fast events occur shows that time is measured by changes and changes are measured by time. Patterns and their accumulation give two conceptions: cyclic time (e.g., seasonal) and linear time (e.g., the unrepeated irreversible process of decay and death). Which conception of time is emphasized depends upon our interests and values.

These general remarks concern our everyday experiences. When we turn to science and mysticism problems arise. For example, the unidirectionality of time has been challenged by a theory in quantum physics that views positrons as electrons flowing backward in time. Other quantum problems have given rise to the issue of whether time is continuous or quantized. Relativity rejects the Newtonian notion of absolute time (i.e., time as an independently existing component). This Newtonian concept, coupled with entropy from classical mechanics, seemingly gave credence to our everyday views of time. With relativity, problems arose with the nature of simultaneity, that is, the idea of events as occurring simultaneously regardless of the frame of measurement. The result is the theory in which there is no absolute "now." In different lines of succession, within very strict limits, event *a* may appear to occur prior to, simultaneously with, or after event *b* depending upon one's frame of reference. Time becomes route-dependent, and no measuring system has a claim to absoluteness.

Time in relativity theory is open to two philosophical interpretations, one involving "becoming" and the other "being."[23] The former is more familiar, and more popular among scientists: change makes up a genuine component of reality. For the latter, events do not *happen;* we just come across them. They are there independent of us and we become aware of them. It is a static view of the world of phenomena. From Hermann Weyl: "Only the consciousness that passes on in one portion of this world experiences the detached piece which comes to meet it, and passes behind it as history, that is as a process that is going forward in time and takes place in space."[24] There is no "flow" of time in addition to the real events. Some would justify one idea of time-travel with this: our consciousness at least can move forward and backward to different spots of the timeless whole.[25] Reality is a timeless, changeless whole—clocks do not measure anything real, nor is there an objective past/present/future sequence. Relations of "before" and "after" remain unchanged, but the "flow" of time is denied—and change and all relations of "cause" and "effect" also become illusory. Events become like still frames of a movie with our consciousness providing the action.

The first interpretation does not warrant more comment. But the second—that time is not a component of reality—seems *prima facie* to be relevant to those nature- and depth-mystical systems which also advocate a static view. Time becomes a religious concern with the realization of changes important to our lives—especially births and deaths. The suffering endured throughout a lifetime requires an explanation to make it acceptable. It must be shown to have a rightful place in the scheme of things. Myths, among their diverse functions, provide such explanations. Within Indian culture, this mythic response is present, but of interest here is the occasional emphasis upon impermanence and suffering to the point of denying the ultimate reality of time. The beliefs in rebirth and in the huge number of world-eons causes time to be equated with death or a destroyer (e.g., as in *Mahābhārata* I.1.273 or *Bhagavad-gītā* XI.32). Time becomes the source of death, not life; only by getting out of time do we become truly "alive" (real).[26] Yet it is not time itself that "fetters" us: the illusion of *māyā* is taking the objects appearing in time as ultimately real and thus to believe that nothing exists outside the realm of time. "One is devoured by Time, not because one lives in Time, but because one believes in its *reality*, and therefore forgets or despises eternity."[27] Freedom is then living spontaneously, knowing the true nature of things.

With regard to nature-mysticism specifically, the being-ness of the whole is "outside of time" or "eternal" in the sense of lasting throughout all periods of time. Changes occur to the impermanent entities within the whole, but the whole is eternal and thus changeless. But such a denial of time differs radically from that within the static interpretation of relativity. Neither science nor mysticism denies the order of before and after: within any relativistic frame of reference the order is not variable but fixed; nor do the Theravādins (or the Advaitins) affirm that we can set any temporal sequence we desire. Nevertheless, for the Theravādins change (impermanence) is part of the true nature of reality, precisely what is denied in this interpretation of relativity. The idea of consciousness as passing over changeless (static) slices of the world's history goes directly counter to this; in fact, the reality of the entities and events would be supported by this interpretation of relativity: things would be changeless, eternal and immutable rather than being void of substance and impermanent.

Thus science and nature-mysticism have little in common in this matter. Time is not a separate topic in the Pāli *Nikāyas*, but the idea of time in its everyday sense of a continuum of past, present, and future is assumed. The Buddha did dismiss the question of the possible infinity of the world in time as irrelevant to the religious problem (although elsewhere he states that the cycle of rebirths is beginningless). The interest is upon present action, but *kamma* and rebirth create interest in the future. Past actions can bear fruit in the present or distant future. Thus events in the past have real effects afterward. We can also remember our own past lives, past lives of others, and the course upon which *kamma* will propel people by means of paranormal powers *(iddhī).*[28]

For thė transition to the issue of time and depth-mysticism, timelessness and eternity, and two types of timelessness must be distinquished. *Eternity* is the infinite extension of temporality into both the past and the future; the *timeless* is something totally outside any temporal framework, something to which temporal categories do not even appear to have application. Within nature-mysticism, mindfulness involves, in its extreme form, a total lack of temporal awareness, that is, no ordering of the flux of one's experiences within a temporal framework. We live then solely in the present; the past and future do not enter into consideration. Any awareness of the experience as placed in a temporal framework (either of duration or before/after) would interfere with this experience: awareness of time would be a reflection that would destroy the complete attention paid to what is immediately at hand.

Thus it is said that for the enlightened, "there is neither past nor future" (S I.141). Yet in less extreme forms of nature-mystical experiences, such as enlightened states of mind, concepts still structure sensory awareness to a degree and the future consequences of actions enter into consideration. Time thus becomes part of the experiential framework. The permanence of the mindful enlightened state and the end of the cycle of rebirths, however, permits the Theravādins to speak of *nibbāna* as being outside the realm of change and thus outside of time. It is the not-born, the not-become *(abhūtam),* the not-made *(akatam),* and the not-constructed *(asaṅkhāram)* (*Udāna* 80). The Buddha is one who has "transcended the world-ages"—he is not a "man of the ages" (*Sutta-nipāta* 373,860).

The timelessness of the depth-mystical differs. There is still the need for a total lack of a temporal awareness during the experience, but in addition there is now an awareness of a reality outside of time, that is, a reality that is timeless as opposed to eternal. In nature-mystical experiences, the awareness may involve no experience of time, but temporal categories are nevertheless applicable to what is experienced (the flux of parts within the eternal, changeless whole).[29] The "object" of the depth-mystical experience (the mystical), however, is changeless in every way and thus outside of time in a more radical way. Śaṁkara does not make time a major topic but does say that reality (Brahman) is without change and that past, present, and future do not apply to it (BSB I.1.14). Change and the entire before/after relationship do not designate anything of the true nature of reality, according to such depth-mystical systems. All before/after ordering is no more indicative of reality than that the sun's crossing the sky indicates that the sun moves. Temporal relationships are an inherent part of the phenomenal realm, but this realm has the status of a dream: its contents, including time, are ultimately not real.

Misunderstandings might arise when mystics speak of "now" or "the present" as the real. The present is the period in which experiences actually occur—this is what distinguishes the present from the past and future. According to relativity, there is no absolute now, no universal present moment. In ordinary usage, "now" refers either to the time in which experiences occur or the datable point within the temporal framework whose length depends upon the context of the discussion—"now" may mean this second, this year, this historical age, or this geological age. Mystics subscribing to a nondualism speak of "now" as the totality of reality because of the belief that temporal succession is only a shadow-play, not reality. All entities are unreal in the final analysis; thus they are not present in this location at any given moment, but the reality constituting any phenomenon is present—and since this reality is indivisible it is totally present in each phenomenon and each "now." Thus each "now" is the sum of reality: because all reality is ultimately nondual, whenever any phenomenon is, the reality of everything is. And so too each experiential "now" is exactly the same. Thus the mystical is manifest only in the present, and the present only in the atemporal experiential sense, not that of a moment separating past and future on the continuum of time. The temporal "now" of today is no nearer to or farther from the mystical "now" than the "now" of a million years ago, since the experiential present—the access to the mystical—is not a movement on the temporal continuum. The experiential present is outside the continuum of time; when any moment "moves" past it into the past, it does not change. So too the mystical is not "in" time; it is the "eternal now" not open to temporal categories but available completely and changelessly at any time.

During mystical experiences, the mystic is both in time and outside of time. The question of the experience alone is easily resolved:[30] an experience of the timeless mystical, like any experience, must take place in an interval of time (as measured by anyone); only in this case this interval is not sensed (the sense of temporal categories would destroy either the depth- or nature-mystical experience). But the homogeneity of clock-time remains the same regardless of what experiences are occurring. Thus the enlightened may not employ time as a category for organizing their experiences, but to observers they are objects in time; and obviously the enlightened still grow old and die.

Because relativity under both the dynamic and static interpretations is concerned with the *measurement* of time (clock-time) while mysticism involves a different type of *experience* of reality, the two cannot be related easily. Traveling at speeds close to that of light, we would not experience time differently—our clocks would not seem to us to be moving at a different rate. At those speeds, from our point of view our "lifeline" is always the same; but from the point of view of an observer in another frame of reference, our "internal clock" (e.g., heartbeat rate) will have slowed down and thus we live longer (as illustrated by the well-known twin paradox). Similarly, the denial of an absolute "now" in relativity means only that there is no absolute temporal framework (i.e., one "correct" for everyone regardless of motion) with which to measure time. Even the static timelessness of one interpretation of relativity still has diversity in its still pictures and thus is not the mystical real; in such a view, it is difficult to speak of a "now" in any meaningful sense at all. None of this can be related to the lack of awareness of time's passage in mystical experiences or to the reality allegedly experienced.

There are other senses of timelessness within science. Mathematical propositions are timeless in the sense that the equation "$2+2=4$" has no reference to time. And most sciences involve mostly historyless laws, that is, there are conditionals with time as a parameter, but they are true of any interval of time. But this does not negate the underlying importance of time to the sciences: science involves relationships that remain constant throughout the process of change. While mysticism is interested in the present only, science's interest is in the order occurring throughout the entire past/present/future continuum. If anything, the sciences are not interested in the present at all.

Space

Before relativity, the nature of space was debated in the modern West (e.g., the Leibniz-Clark letters), but the Newtonian conception was the most usual: a big empty "box" without any sides or top or bottom in which objects are placed. It was conceived as absolute, that is, there were fixed points "of" space "in" the box. With relativity, space lost its absoluteness. It became integrated with time into a new fundamental reality: space-time. Neither one has independent reality in this conception, but the spatiotemporal continuum is absolute, that is, physically real and "independent in its physical properties, having a physical effect, but not itself influenced by physical conditions."[31] The continuum is not empty but a "field" (modeled on a magnetic field) affecting matter in it—or, according to another interpretation, matter is just disturbances in space-time. The "curvature" of the continuum depends upon the amount of matter in a region; it is not uniformly "flat" (Euclidean). Curves, not Euclidean straight lines, are the shortest distance between points.

Space in relativity theory is not uniform and static as our normal experience would suggest. But its dynamics differ from the general experience of sacred as opposed to profane space, that is, certain locations as having power in themselves (at least during rituals) that other spaces do not and as providing fixed points of orientation related to the meaning of a total way of life. Relativistic space has the uniformity of being strictly law-governed, even if it does not possess the Euclidean type of homogeneity. So also the scientific studies of space relate only to its physical level of significance—the experience of it for scientific purposes is uniformly mundane, not broken up into discontinuous areas of unimportance and special importance. Nor does our *experience* alter with the length-contraction occurring measurably at great speeds. The architecture of space is studied for problems comparable to those of clock-time, not to our experiential reactions of how fast time passses.

For nature-mystical systems, space, like time, is not a major concern. The being-ness of phenomena, not their spatial arrangement, is central to the experiences themselves and the mystical systems in general. The theory that time and space each arose with the origin of the universe in a "big bang" (i.e., before that event, there was no change and so no time, and no extension for spatial diversity) may be integratable into nature-mystical theories of the whole as the source of the parts. So too space-time as a unity may be reconcilable to the claim that no parts of the phenomenal realm are isolatable and real. But there is no thought in classical Indian texts in this direction. Nothing suggests that space *(ākāśa)*[32] is dynamic or affects objects, or is ontologically the same as they are by being the reality out of which objects arise. Nor are space and time connected ontologically; they are not even discussed together (space getting discussed more than time). In no way is space an especially important reality. Still, in the nature-mystical experiences, all phenomena are not contracted into a point: there is extension even though the experiencer is not using spatial categories in structuring experiences. The Theravāda texts seldom refer to space, since it is not even tangentially connected to their religious problem. When it is discussed, the common Indian beliefs are affirmed.[33] Thus space is the element upon which other "great elements" *(mahābhūtā:* earth, water, and air—fire is not mentioned here) rest (D II.107), while it itself does not rest upon anything (M I.424).[34] This is not exactly the idea of an empty receptacle into which objects are placed, but *ākāsa* is considered "unconstructed" or "not-conditioned" *(asaṅkhāra,* not born of *kamma)* (Mlp 268)—a property otherwise reserved for *nibbāna.* This indicates as strongly as the Theravāda tradition can space's unchangeableness and independence, features uncharacteristic of the Theravāda and other nature-mystical discussions of reality in general. This is not a conservation-law for the content of the flux within space, but a claim that *ākāsa,* like Newtonian space, is absolute.

For depth-mysticism, the ultimately real transcends any spatial distinctions. Brahman is beyond space (BU IV.4.20); space is woven across the imperishable (BU III.8.8). That is, no spatial terms apply to Brahman, and the extended phenomenal realm is a "covering" of the real. The same problems with "now" with regard to time arise here with "here" and space.

The image of the mystical as the "still center" around which everything whirls is not applicable to the Advaitic conception: the spatial and spaceless mystical cannot be represented as two parts of one whole. The mystical reality is nowhere among sensory phenomena—it is extensionless, not ubiquitous—but all reality is present in the depth "here." Each "here" is capable of revealing the mystical, that is, each point contains all the power of reality. In fact, according to Advaita, the sum of the indivisible reality is in each "here." Nothing related to the measuring systems in science bears upon the experiential claim that each "here" is the center of reality.

Orderliness and Connections in Reality

Another area that illuminates the nature of science and mysticism is their respective claims on the orderliness, and mechanisms accounting for their orderliness, among natural phenomena. Order is revealed in the sequence of events if predictability is found. In science and everyday experience, this is manifested most usually by causation. "Same causes, same effects" is the key for scientific order and connections. Since the sciences are interested in the mechanisms of change, it is the Aristotelian efficient causes (that by which change is brought about) that are of central interest. This is true at present in the physical sciences, although teleological causes are still debated for biology. It should be noted that there is less emphasis upon causes as science moves away from everyday phenomena, the context from which the concept arose, into theoretical understanding. But causal order is important enough to suffice for this discussion.

As with space and time, it is easier to identify specific causes than to analyze the concept. Problems appear from many quarters: the Humean objection that all we actually see is the constant conjunction of two types of events (we supply by force of habit the ncessary connecting tie between them); the matter of how to define "similar conditions" in a noncircular manner; the issue of why causes (the temporally prior event or more substantial entity) explain effects (the following or subsidiary event) rather than vice versa. Causation may in fact only give the illusion that we understand nature. It is true that causation is not necessary to our experience—we can experience the world in a noncausal manner: the extreme instance of nature-mystical experiences involves not seeing any temporal connections or dependence-relations but merely the pure flux of phenomena. Therefore we can conclude at least that causation is a regulative principle connected to a type of learned experience and that causal discourse is theory-laden. So too as part of the conceptual organization of experience and theory, what is deemed a cause becomes dependent upon the science in question and the specific theory: "causes" are placed into wider explanatory schemes dependent upon considerations other than mere observed correlations.[35]

Causal connections do involve more than merely constant succession or predictability: day follows night, autumn summer, without our being induced to speak of causation as being involved. So too some conjunctions of events appear to be mere coincidence. When causation is involved, we feel that something in the workings of reality beneath surface-appearances (the observed constant correlations) brings about the effect: a hidden "power" or necessary connection is involved. In science the cause is not the sum of all conditions surrounding an event. Rather it is the *sufficient* or *necessary and sufficient* conditions to bring about a change. Background conditions (i.e., those which are necessary but which remain constant throughout the change) are not designated "causes," because scientists are interested in the change only. Thus it is assumed that, although perhaps every object in the universe may affect any one particular event, some conditions precipitate a change as an efficient cause and some do not.[36] Singling out one condition as *the* cause (as in the analogy of a causal chain) ignores the other relevant conditions.

Causal laws are in the form of conditionals: whenever conditions x are all fulfilled, phenomenon y appears. This correlation of x and y is observed regardless of whether we are correct in thinking there is something more behind it. Predictability approaching certainty is a feature of causation in our everyday realm of experience. In the submicroscopic realm, lesser probabilities are the best predictions attainable.[37]

The only necessity of causal laws is simply that whenever the initial conditions are satisfied, the effect inevitably follows according to the way described in the conditional. The conditions may never again be fulfilled, but this does not affect the law. Causal conditionals differ from *determinism* in that the latter deals also with the fulfillment of the initial conditions: one effect is sufficient for another to occur, and so on in a self-perpetuating manner. Causal laws do not deal with how or whether conditions are in fact ever satisfied. Using the paradigmatic instance of the billiards game, causal conditionals are of this sort: given the background conditions (laws of mechanics, gravity, the coefficients of friction between each ball and the felt, etc.) and the initial conditions (the position of the balls), if the player shoots the cue ball in such and such a manner, a certain final position of the balls will occur; after each shot, new conditionals will be set. But the player is free to choose what shot will be taken at each juncture. For determinism, the entire course of the game is established from the background and initial conditions (perhaps including the psychological make-up of the player). Causal conditionals show what would happen *if* a particular shot were chosen, not how the game *must* proceed. Thus determinism goes beyond the minimal article of faith in the uniformity of nature necessary for scientific understanding. Problems with determinism (e.g., the probabilities of atomic phenomena and the theory that since the big bang the universe becomes increasingly more complex so that information available at any point in time, even if complete, is insufficient to act as a basis for accurate predictions concerning the whole of the universe) do not affect causation. Nor does the orderliness of causation rule out miracles in the original sense of any event's manifesting the action of God regardless of causal status, or in the more modern sense of an event's violating a law of nature. Causation involves conditionals only, not naturalistic metaphysics, and miracles (if there are such things) would merely keep the initial conditions of a conditional from being fulfilled. The metaphysical belief that all events have natural (i.e., law-governed) causes is not part of causation *per se,* although it is part of the metaphysical articles of faith of science.

Other possible objections to causal laws do not hold up. One comes from the relativity of frameworks, but causal order is invariant through change of frames of reference. Another possible objection: within everyday experience, different cultures may emphasize different aspects of the causal sequence or ignore the category of causal connections entirely. Nevertheless, causation is not a purely social product in the sense that we could set up any temporal or causal ordering we chose. Even the acausal orderliness of Jung's and Pauli's synchronity (meaningful coincidences occurring without causal explanation) would not be counterevidence against conventional causal conditionals. The fact that laws may not apply, say, in black holes may count against determinism, but if the antecedent of the conditionals arising from more mundane situations is not fulfilled, causation is untouched. All laws have limitations specified by the initial conditions. Further research would count against causal laws only if the same conditions produced different results in other parts of the universe. Laws of nature may not be immutable.[38] This would greatly restrict the range of laws but not vitiate totally their usefulness and applicability.

Turning to the Indian traditions, we see that here also there is much emphasis upon order. There is very little in the way of teleological explanations, for example, that the universe is evolving toward some goal, as with Teilhard de Chardin. But in the Vedas, *ṛta* is an order controlling nature, society, and even the gods. It is inviolable with regard to nature, but part of Varuṇa's tricky power is to make the moral aspect hard to know. *Ṛta* is the source of harmony between parts of the universe; it is not a power like Brahman, but more like a law describing the truths of the universe. It does not exactly fit our category of a "natural" order or connection because, although it is not created by Varuṇa or the other gods, it is guarded and enforced by them. K. N. Jayatilleke is correct in saying that there is no causal order in the *Ṛg-veda:* the power is not in a connected system of events governed by laws, but in the wills governed by the order *(ṛta)* of the gods behind phenomena.[39]

The Theravādins and Advaitins accept absolutely an order among phenomena. Śaṁkara, in accepting such order, says that consciousness, not nonintelligent matter, is the necessary cause of it (BSB I.1.15). The early

Theravādins accepted change among phenomena as real, not illusory, and accepted continuity of such change (the doctrine of momentariness [*kṣana-vāda*], with its problems of accounting for continuity, coming only later). All changes are law-governed; there is no randomness, for that would curtail our ability to guide our life toward *nibbāna*. The Theravādins defend continuity without fatalism (determinism); this permits freedom to elect courses of action, and guarantees control through predictability.[40] The orderliness ("that being the case, this arises") is not our superimposed habit of thought. There is a nature of things *(dhammatā)* and an "objectivity" (*tathatā*, "that-ness") of relationships that exists independently of their discovery by the Tathāgatas (S II.25–26). The permanence of the relationships does not necessarily go against the Buddhist claim that all constructed things (i.e., what is governed by the relationships) are impermanent and without substance. That is, the principle of change and the more specific laws of *kamma* and dependent origination are themselves changeless. So also a conservation-principle for the totality of reality would not invalidate the claim that all is in a state of flux. The issue (not discussed by the Buddha) reduces to whether laws and principles are within the category of "constructed things." In any case, the order to the flux that such laws impute is what is of importance here.

How do the Theravāda concepts of order and connections compare and contrast with those of science? More specifically, are the connections between phenomena comparable to scientific causation? Often it is said that mystics, like scientists, search for the causes; but all this may actually mean is that they too search for the reality behind appearances, not that any mystics have a commensurable sense of "causation." The Theravādins accept the conditionality of all phenomena, search for the conditions involved in change, and note some regularities of relationships related to the religious quest. The relationships are as natural as those of science, that is, no agent or force not bound by a law intervenes. So too predictions in their lives are made based upon these regularities. But not all orderliness is necessarily causal; only connections lawfully *producing* an effect are.

Two instances of orderliness and connection are often cited as instances of causation in Theravāda Buddhism. First, *kamma*. "*Karma*" in Indian thought covers both the deed and its consequences, which lawfully accrue to the actor. For the Theravādins, our present and past deeds (and our thoughts at the time of death) play a role in determining the future course of events in an orderly manner. The kammic results are not a mechanical product of the act alone, but depend on the *intention (cetanā)* of the actor, the circumstances of the act, the spiritual worth of any person impinged upon, and other conditions; this complicates the picture, but does not weaken the lawfulness involved. General correlations of acts (*kamma* in the restricted sense) and its fruit *(phala)* can be illustrated as follows: a killer will be short-lived, a nonkiller long-lived; the miserly will be poor, the generous rich (M III.203f). Another example: the Buddha's disciple Moggallāna in his last life had to be murdered (because he murdered his parents in a previous life), and the robbers who did it also had to do a deed worthy of execution because of their previous deeds. Such interconnection of different *kammic* paths are commonly recounted. But not all events are determined by past actions: the effects of the past are not new causes in a deterministic chain.[41] Nor is *kamma* the only source of effects (Mlp 136–37). We are free to choose, and the results of our volitional action will have future effects.[42] That *kamma* is lawful is no more damaging to "free will" than gravity: our choices might be limited but not determined.[43] Moggalāna could choose the time and place of his death: how he died was fixed, but there was freedom within the boundary conditions. That is, he could not affect the mechanics of the lawful order but could modify the application-conditions. Kammic fruit also is not inevitable in every instance: minor evil fruit can be eliminated by meditative practice, although major evil fruit (such as Moggallāna's) cannot (Mlp 188–89). That is, the kammic order is conditional upon the fulfillment of certain conditions, and these conditions will not be satisfied if meditative interference occurs. But the destruction of fruit also proceeds only in an orderly manner, not capriciously, depending upon the meditative practice.

Kamma does appear from this to satisfy the conditions of causal connections between phenomena: there is a natural order, and the deeds are the sufficient conditions for the fruit to occur (if there is no intermediate interference). It also involves a certain construal of experience: we do not see the connections of the actions and fruits on the surface. Eliot Deutsch refers to *karma* as a "convenient fiction," which is undemonstrable but useful in interpreting experience.[44] Although an instrumentalist theory of terms is no more successful here than in science, *kamma* and rebirth can be profitably seen as theoretical principles that make more sense out of certain experiences for the Buddhists than alternatives (e.g., complicated paranormal powers to explain alleged remembrances of past lives). Most important, it explains evil: some events of suffering that appear as accidents of birth or fortune to Westerners seem natural and law-abiding to Buddhists and other adherents of *karma*. *Kamma* is a mechanism, like gravity, necessary for understanding and also for predicting future occurrences. And like other explanatory concepts, exactly how it works is difficult to specify. Some Indian traditions thought a conscious agent (God) or an "unseen force" *(adṛṣta)* was necessary to make *karma* work; other traditions thought it was self-operating or needed no explanation. When later Theravādins did explicate more of the conceptual framework, problems arose. In particular, as Ninian Smart says, *kamma* is "action at a temporal distance" and the doctrine of momentariness makes this hard to understand.[45] If all the factors of the experienced world arise and fall in each moment, kammic efficacy would be cut off each moment and thus no kammic fruits could accrue to acts.

Is *kamma* strictly scientific? That it regulates something "moral" or "conscious" rather than "physical" does not make it any less natural or lawful. In addition, according to Indian belief, our volition affects not only individual but also social and even physical events, the social and physical orders forming one whole. In general, our distinction between "consciousness" and "matter" is not important to all people. But a basic problem related to predictive capacity of *kamma* is produced by its relation to rebirth. Most people in our culture would dispute rebirth as the most plausible theory because of its conflict with other, more deeply held beliefs. But if we accept its reality, the Buddhist advocates of *kamma* can always say that if a kammic prediction goes unfulfilled, some past act is responsible for its occurrence (since the chain of rebirths is indefinitely long, according to Buddhists). Thus the empirical claims cannot be disconfirmed. And advocates of other theories would explain confirmations as due to other reasons or as coincidences. Thus *kamma* may have too many problems with falsifiability to be clearly scientific rather than metaphysical; but it is still causal.

If *kamma* is causal, what about the relationships of phenomena summarized in the Theravāda formula "This being, that arises; this not being, that does not arise"? The Theravādins accept the doctrine that the consequent is not the same as the antecedent condition described in the conditional (the *asatkārya-vāda*). The multiple conditions come together with a result occurring that is not identical to the conditions, that is, the result does not "pre-exist" in the conditions. Thus there is a real change, not a mere appearance of change. The conditions and result are neither the same nor different, that is, they are not identical nor totally distinct but related; the result does not emerge *ex nihilo* or fortuitously but is a well-ordered, genuine change.

All the factors of our experienced world (the *dhammā*) are so connected. The most important instance of this dependency from the Buddhist religious point of view is its central prescriptive guide, dependent origination. If each condition in this formula is necessary and sufficient for the result to arise (nescience for dispositions, and so on), as the abstract dependency-formula suggests, and if the formula is meant to be truly circular (nescience eventually gives rise to craving, which eventually gives rise to more nescience, and so on), then an unbreakable chain has been established—a *determinism* of causes and initial conditions. Once unenlightened, we will always be unenlightened. But if we look at the actual conditionals, we see that certain of them state merely the *necessary conditions* for the arising of the subsequent element.[46] The important difference is that the necessary conditions do not produce a deterministic chain but can be eliminated, thereby breaking the cycle of rebirths. Some elements in the formula are necessary and sufficient for the arising of the subequent element (e.g., craving produces a new rebirth). But not all of the elements are such necessary and sufficient *causes.*[47] For example, death is necessary for a rebirth to occur, but it is not sufficient (the enlightened are not reborn). So too the Buddha had feelings (D II.157), but these feelings did not give rise to craving. For the elements of the formula, *x* does not always produce *y*, but if *x* is absent, *y* cannot arise. The necessary condition of nescience can be eliminated, thereby ending craving and thereby ending the process of rebirth. Thus nescience begins the formula because of its logically unique and fundamental role in the Buddhist conceptual scheme. Thus too, unlike the abstract dependency-formula, dependent origination makes the religious life possible and depicts how to end rebirth by cutting off nescience and craving.

As with *kamma,* there is no absolute division here between the mental and the physical, each supplying conditions for the other. And like *kamma,* dependent origination is an explanatory mechanism making the flux of certain experiences intelligible and predictable. But unlike *kamma,* dependent origination is not an instance of causation: its conditionality involves a mixture of prominent necessary conditions and some necessary and sufficient conditions (with nescience fulfilling a special role), not the exclusively necessary and sufficient or alone sufficient conditions of causation. Commentators describe it as causation, perhaps to make it sound more scientific than it is.[48] Nevertheless, as Kenneth Inada points out, dependent origination is not the traditional Western problem of causation but something unique to the Buddhist tradition.[49]

When we turn to Advaita Vedānta and the question of causation, we see a major shift accompanying Advaita's emphasis upon being. Śaṁkara repeatedly claims that Brahman is the only "cause" *(kāraṇa)* involved in reality: it is the substantial *(upadāna),* efficient *(nimitta),* and activating *(kartṛtva)* cause of whatever is (BSB II.3.35). It is the total substance and ruling principle of everything (I.4.23). No supplementary power or aid is needed (II.1.24). Brahman is responsible for the absolute orderliness of phenomena—no structuring element independent of the substance is involved. For this an all-powerful and all-knowing consciousness is required (I.1.2); the unconscious matter *(prakṛti/pradhāna)* of the Sāṁkhya will not explain the order, according to Advaita.

The problem here is that causation normally means what brings about a change in the phenomenal realm. "Cause" and "effect" are both on the same level of reality. A cause is a component of the realm of space-time, not of the dimensionless world of the mystical. Connections and uniformity may be present in both science and nature-mysticism, but with depth-mysticism it makes sense to speak only of unity (nonduality). Scientific causation is a matter of "horizontal" connective orderliness, while mystical "causation" is a matter of "vertical" unity, the "depth" of reality. If the scientific levels of reality provide the necessary and sufficient conditions for the surface-level (so that it would make sense to speak of *causation* between levels of reality), still the mystical reality as portrayed in Śaṁkara's Vedānta does not have the capacity to be a scientific explanatory entity (as discussed earlier): what causes everything is not what we can consider a cause of anything in particular. Conversely, the mystical is beyond all karmic and other causal relationships: ordering is within the phenomenal realm and Brahman is outside it.

The "power of being," as opposed to a force of nature, can be referred to as a "cause," even the primary cause instead of secondary (scientific) causes, since it does provide sufficient conditions for the phenomenal realm as a whole, if not particular phenomena, to appear. Again, this "first cause" is not first in the horizontal line of causes and effects (since there is no change occurring). The depth-cause also has a type of "omnipotence": the mystical "controls" events by supplying their being in a timeless "creation." The mystical does not enter the surface-causal processes (science may be able to give an exhaustive explanation of these processes) but is neutral to all action and individuality. The mystical makes no difference in our understanding of the phenomenal facts; hence the philosophical problem of proving the mystical exists from data within the "dream."

Another problem arises in referring to the nontemporal reality behind phenomenal relationships as a cause. From the point of view of nescience, it makes sense to speak of the real as containing or causing the phenomena; but from the point of view of knowledge, the phenomena *are* Brahman merely mis-seen. Whatever reality there is "in" phenomena simply is Brahman. Thus there is no *action* by Brahman: it is not a cause *in* the world or *of* it—hence the doctrine that within the phenomenal realm causes and effects are nondifferent (the *satkārya-vāda).* The "being" of the cause and effect is the same, but the two appear to differ (and genuinely do differ with the phenomenal realm)—thus the label "nondual" rather than "the same." And since "being" is the only criterion of reality recognized by Advaita, no real change or production occurs. There is an emanationist theory within the phenomenal realm—minerals are "modifications" of earth (BSB II.1.23), water is produced from fire (BSB II.3.11), and so forth. And there is absolute order within this realm (e.g., BSB II.1.27). But since Brahman alone is real, all reality is the same; all variation is in the final analysis mere appearance, not a transformation of Brahman (the *vivarta* type of *satkārya-vāda).* We take variations as real only because of nescience. In reality there is no "causation" between being and phenomenal appearances. Nothing happens—no real change or occurrence is produced. Causation reduces to just our way of ordering events in the shadow-realm. Advaita may appear to have this last point in common with the static interpretation of relativity. But the differences are significant: relativity, while denying the flow of time and causation, is dualistic in one regard (consciousness and material objects) and pluralistic in another (multiple objects), and affirms the order among objects as real; there is nothing about a depth-being constituting the reality of everything.

The difficulty here with Śaṁkara's position corresponds to the problem of how the "one" can be the ground of the "many." How does the changeless "produce" the realm of change? How does the unsensed "become" the sensed? How can the "dreamer" remain unchanged in producing the "dream"? For Śaṁkara, Brahman needs no instrumentation for producing the world (BSB II.1.24–25); and since the emphasis is on "being," the issue of explaining the mechanics would be unimportant to Advaitins. Brahman is also responsible for the lawfulness of *karma* and other order: it provides any hidden connections between "cause" and "effect" in the phenomenal realm (if there is any necessary connection). Would this not mean that Brahman has structure in addition to a power, even if Brahman is not compelled to "create" in a lawful way?

Paranormal Phenomena and Order in Reality

Another aspect of the Indian traditions has importance for the issue of the lawfulness of reality. It is those paranormal powers *(iddhī/siddhī)* which allegedly can affect matter. These powers are not directly related to mystical enlightenment but are associated with meditative progress and are part of all mystical traditions. Mystical knowledge does not give control over phenomena: it involves how we see reality, not interfering with, or affecting, reality. The Buddha and others often warn against too much interest in paranormal powers, since they may prove to be obstacles on the path to enlightenment: such powers may increase our sense of self-importance or facilitate our attachment to worldly desires.[50] The only use the Buddha is reported to have made of these powers is to convert people to his way of life.

The Theravādins discuss these powers more than do the Advaitins. The powers are one of the concentration-produced higher knowledges *(abhiññā)*. They include making oneself multiple or invisible, traveling through walls, sitting cross-legged journeying through the sky, touching the sun and the moon with one's hand, and reaching Brahma's world while remaining in one's body (D I.78). The effect of mind upon body is more readily noticed in yogic control of involuntary bodily processes (e.g., changing the heart-beat rate or the temperature of parts of the body). Classical Indian texts contain grander claims, the division of mental and physical not being of major importance. The orderliness of society affects the orderliness of natural processes (A II.74–76), and the disruption of society may produce floods, droughts, or eclipses. The *Kurudhamma Jātaka* tells of famines and plagues produced as a kammic effect of a king's not observing the basic lay Buddhist precepts. So too, during meditative trances, one is impervious to fire, poisons, and weapons. Even words in certain circumstances can have this power: by the act of truth *(satya-kriyā)* one can accomplish such feats as making the Ganges flow backward (Mlp 120–23).

More radically, every hell and heaven is literally man-made by persons with similar kammic destiny; each will disappear when it is no longer "occupied" by anyone. Does our "physical realm" *(kāma-dhātu)* differ from these other realms in this regard?[51] Any radical dependence of the physical on the mental would be hard to reconcile with the Theravādins' more usual realism concerning laws and factors of the experienced world.[52] Even the Yogācārins, who emphasize the role of mind and who advance, for want of a better phrase, a "collective subjectivism" ("storehouse consciousness," *ālaya-vijñāna*) to account for the regularities and other features common to our experience, still posit a reality existing independently of our individual minds. From a contemporary Western point of view, there is a sense in which any culture "creates" the social and even the whole phenomenal world by emphasizing only certain aspects of flux of reality existing apart from us; we make the "facts." But such adjustments within an already-existing and independent reality differ fundamentally in conception from any idea that an entire realm of reality depends ontologically upon the mind.

Scientists have a general reluctance to accept even the limited claims of mental control of matter; such claims conflict with more deeply held beliefs, are difficult to confirm in the analytic conditions of a laboratory, and so forth. Everyday interaction (if that is the correct relation) of "mind" and "body" in such instances as raising one's arm or as when emotions produce physiological effects is readily accepted. And there is a theory (albeit not a widely accepted one) employing consciousness as the "hidden variable" responsible for the actualization of one of many possible events for certain submicroscopic phenomena. If certain exotic macroscopic phenomena such as walking barefoot upon glowing hot coals without burning one's feet or clothing become well-attested, the problem will become the explanation of the phenomenon, and not its reality.

One area of opposition to such phenomena can be removed. That is, paranormal phenomena can be interpreted so as to provide no counterevidence to the idea that there are fixed laws of nature. Let us assume that consciousness can be a type of "energy" that can enter physical events, or at least that somehow consciousness can affect matter in instances other than normal bodily ones. In normal "states of consciousness" (the mind organized to function in what we accept as normal activities), we assume that our observational awareness does not affect what is observed in the macroscopic world. Physical laws are the results of such observations; that is, lawful conditionals hold when the human mind is not a causal factor in events. When the mind does enter into events through altered states of awareness (a change in the operational structure of the mind, not a mere change in beliefs accepted), the antecedent conditions do not obtain, and therefore the law (i.e, the conditional) is not violated. Something outside the province of the law happens instead. The range of the conditions that are necessary and sufficient for a law to hold may have to be revised to include reference to states of consciousness. Surface-phenomena may be more complex but still have laws definable from reflection upon paranormal experiences. Or there may be no fixed reality apart from our stance toward it. Lawful reality may be literally a product of a particular state of consciousness, and science may be restricted to that state. Still, such an extreme alternative need not be required for the acceptance of the idea that paranormal events may occur that do not invalidate the lawfulness of everyday reality.

5
The Nature of Knowledge

Scientific and Mystical Knowledge

Turning to the issue of epistemology (the nature of what is accepted as knowledge and the justification of knowledge-claims), we see that "knowledge" and "reality" are interconnected concepts: what we accept as known depends upon what we accept as real. "Truth" too embraces both. A central Sanskrit term for reality, *"satya,"* in fact means both reality and truth.

Truth and knowledge in the philosophical analyses inspired by science as the source of knowledge are taken to be propositional in nature: a statement is true if it asserts what is the case, and knowledge is a justified belief in what is true. Such knowledge is "discursive," that is, propositions defended by arguments. A simple correspondence-criterion for truth leads to problems of isolating reality from all claims about reality for comparison with those claims. Truth leads to problems involving how we know what is in fact the case, but it does show that we accept assertions as stating something true of reality. Not all knowledge that such and such is the case is linguistic in nature (e.g., animals have such knowledge without any linguistic ability). But all scientific knowledge is conceptual. Even the experiences with which scientific knowledge begins are shaped by such concepts—herein lies one of the strengths of coherence of concepts as a criterion of truth.

In the West, knowledge was once considered to involve certainty: we *know* only what we are logically certain of. For Plato, true knowledge exists only about the unchangeable; in the context of the Greek objectification of thought, only *ideas* fulfilled this timelessness. But such certainty is achieved at best only within mathematics. Within the natural sciences knowledge is more tentative, admitting of degrees of assurance. Scientific revolutions reveal the limitations of our previous "knowledge." Nothing is incorrigible in scientific knowledge.

Mysticism has a different view of truth and knowledge. Mystics do not advance ordinary empirical claims, but they take mystical experiences to be "cognitive" in the sense of providing true insights into the nature of the world. Truth is not merely a statement of what is the case, but is the fundamental, undeniable truths of reality necessary for living the most advantageous life. We align our life with whatever we accept as knowledge, and all advances in knowledge are therefore liberations from the past to a degree. Mystical knowledge differs from scientific knowledge in the radical transformation of our life that is given in enlightenment: it is not a minor adjustment leaving our fundamental values and world-view from before the beginning of the mystical quest untouched. At a minimum, enlightenment transforms our dispositions and attitudes, if not our beliefs. This affects how we live. Religious knowledge in the enlightened state is wisdom *(sophia)*—a know-how concerning how to live, not a mere acceptance of knowledge-that *(gnōsis),* although ways of life do have a belief-component too. The distinction medieval Christian mystics made between knowledge of divine things *(scientia)* and participation in the fullness of God (*sapientia,* "wisdom") is such a distinction. Rather than merely understand and accept a proposition we need to "realize" its truth (i.e., the significance of the proposition must be internalized into our life). The knowledge-claims, even those concerning the mystical, are not so important as a total correct way of living. If an accurate statement were advanced, it would be acceptable—even ultimate truths may be statable—but their acceptance is no substitute for the enlightening knowledge. To contrast with this, the function of scientific knowledge within our culture is that of control of nature and of fulfilling our need to understand reality on one level of significance. The wisdom of how to live and how to use this knowledge is not provided by the sciences themselves.

Vidyā in the *Upaniṣads, jñāna* in Śaṁkara's work, and *vijjā* in Theravāda Buddhism are all concepts of mystical wisdom rather than something strictly comparable to scientific knowledge.[1] The enlightenment-experience gives an immediate "experiential" certainty of the attainment of final truth. In depth-mystical systems such as Advaita, only this ultimately is knowledge, since true knowledge is possible only of what is unchanging. The finality and absoluteness of this knowledge contrasts with science where knowledge is always being revised and ultimately is inexact. Mystical knowledge is seen as completely timeless and historyless: we recover in one insight something always present in our own being rather than discovering something previously unknown by anyone. There are no fresh additions, anomalies, or progress in knowledge once the final enlightenment occurs. Knowing Brahman does not mean merely being able to describe it (assuming some conceptualizations are possible). The necessary knowledge comes only when the mind is free of all ideas rather than in "clear and distinct" ideas.

As usual, nature-mysticism stands between science and depth-mysticism on the issue. In Theravāda Buddhism, unlike Advaita, knowing things "as they really are" *(yathābhūtaṃ)* involves sense-experience and some conceptual structuring; but, unlike scientific knowledge, the emphasis is upon experience rather than an accurate correspondence of statement and fact, although the latter is possible.

The permanence and finality of this knowledge also bear upon the question of omniscience. Science has no key by the knowing of which everything is known (although certain metaphysical positions take science as the sole source of knowledge). The *Upaniṣads,* on the other hand, contain many passages affirming that by knowing *ātman*/Brahman, everything is known (e.g., BU III.7.1, CU VI.1.1). That is, the being of everything is known, although not the particulars of appearance, as in knowing the clay of one clay pot means we know the clay of all. It is "all-knowing" *(sarva-jñā)* in the most important matter of their fundamental nature, not the differentiations and relations of scientific interest. Any facts not related to enlightenment—for example how many moons Jupiter has, or the half-life of uranium 238—are not included. The Buddha had, in addition to enlightenment-knowledge, knowledge through paranormal powers of matters related to *kamma.* But in the *Nikāyas* it is explicitly denied that the Buddha or anyone can be omniscient in the literal sense of seeing and knowing everything at one time.[2] There are passages to the effect that Brahman as creator is omniscient (e.g., BSB I.1.2). If in enlightenment, when we come to know the nature of everything, this all-knowledge of Īśvara is given to us, then the problem arises of how a timeless, unchanging reality (Brahman) could know all temporal facts—if that indeed is the omniscience Brahman as creator has.

Mystical knowledge is sometimes termed "participatory" rather than "dualistic" thought about or observation of what is known. In depth-mysticism, the reality of all phenomena cannot be known as an object, but only as the subject (the witness) (*Upadeśasāhasrī* XV.39–40). In nature-mysticism the separation of object from subject and object from object is broken down. Here to know is to be: to realize our ultimate reality in depth-mysticism, or to realize that we are part of an integrated whole in nature-mysticism. All knowing and understanding break down the opposition of "subject" and "object" to a degree. But the contrast here is that mystical experience and knowledge do not involve the mediation of postulated separate and distinct entities that render sense-experience an experience of a posit-world behind experience and set apart from an observer. Scientific knowledge remains conceptual while mystical knowledge attempts to go beyond this: in science, understanding is (and therefore belief-claims are) the end-product; in mysticism, the experience and the way of life, not abstracted claims, are central. Indeed science, with its generalizations past limited experiences and its inferences to unobservables, is less experiential than mysticism in an important way. Mysticism is "anti-intellectual" in only a limited sense, though, since it claims to give knowledge of reality. But mystical systems are not advanced as intellectual tools to explain experiences or their content. Their explanatory function is only one function involved in the larger concerns of a way of life.

Both mysticism and science accept experiences as necessary, reasoning alone being deemed insufficient. Thus the Buddha criticizes those reasoners *(takkī)* and speculators *(vīmaṃsī)* who base their positions upon reason alone ("who know not, nor see, and are subject to all kinds of cravings") rather than "knowing and seeing" the truth experientially (D I.1f.). Similarly, Śaṁkara asserts that reason alone leads to differing arguments (BSB II.1.11), and stresses the "experiential" nature of enlightenment, although the role of the Vedas as the ultimate source of authority presents problems to any one who feels that mystics take experience alone as the only authority. But mystics do not categorically reject sense-experience and reason. For Advaita, only with regard to the mystical does sacred scripture have absolute authority; in more mundane matters, even *śruti* can be rejected (BUB II.1.20). The *Brahma-sūtra-bhāṣya* is a book of arguments making constant appeal to reason and ordinary observation. Śaṁkara maintains that reasoning *alone* is not enough, but he is not irrational. The Buddha too reasons with his opponents and appeals to ordinary and meditative experiences. For example, we know (according to the Buddhists) that the world is not nonexistent because we see events arise, and is not permanent because we see events pass. In emphasizing experience, mystics understandably degrade concepts, which do not have direct experiential connection. But some understanding must still be provided either implicitly or explicitly. In science, where understanding is central, both reason and experience are more explicitly given weight.

In many ways, mystical knowledge is more like self-awareness than scientific knowledge. Some mystics even refuse to call mystical enlightenment "knowledge," since it could be confused with something more mundane. Rather an "unknowing" of all sensory and conceptual content of the mind is required. Knowing-that presupposes memory for identifications; but mysticism tends toward abolishing knowing-that in order to live in the present. Ordinary knowledge of the mystical, if it were possible, would be comparable to scientific knowledge of consciousness, as opposed to self-awareness. Such objectifying knowledge cannot be substituted for the requisite awareness. From the mystics' perspective, such knowledge would be a matter of images and arguments concerning images.

Concepts and Experiences in Mystical Systems[3]

Other similarities and differences between science and mysticism appear when we compare the relation of concept and experience in the formulation and justification of knowledge-claims in mysticism to that in science. This subject with regard to science was discussed in the first chapter. Do mysticism and science converge in this regard?

It does appear legitimate to speak of reasoning and justifying with regard to mystical experience. To see this, the mystics' own account of the nature of their knowledge must be challenged. Connected to the claim of the finality for mystical knowledge is the claim that one's own tradition has the final truth and all others are either totally wrong or have only part of the answer. The Buddha bluntly states that everyone who thoroughly knows the nature of decay and death, their cause, their cessation, and the path leading to this cessation does so in the same exact way he does (S II.58). For his part, Śaṁkara argues against Buddhists and Sāṁkhya-Yogins. In a text attributed to him, Śaṁkara is said to deny that Sāṁkhya and Yoga lead to liberation.[4] The same certitude carries over to knowledge-claims in maintaining that rebirth and the claim "I am Brahman" require no proof.[5]

To show that all positions within mysticism need supportive arguments, the first step is to realize that mystics are interested in knowledge and the mystical, not in exotic or plesurable *experiences per se.* To say "Mystics are not interested in doctrine"[6] is wrong. The extensive analysis of the *dhammā* in the Theravāda tradition and the arguments of Advaita are related to finding knowledge of what is real in order that we may live the best life. Matters directly related to experiences are part of this, but these and questions connected with how we lead our life, what to expect at death, and so on are all dependent upon how we see the world, i.e., upon the factual component of a way of life. The Theravādins are interested in ending suffering by ending rebirth—their practices make little sense if we do not believe in rebirth. Even if all mystics could concur on goals, values, and world-views, the factual component still could not be ignored.

The opposite error is to give full weight to the conceptual scheme. All experiences are understood in light of beliefs previously developed in a culture, and so it is argued, by Tennant for instance,[7] that religious experiences add nothing to the experiencer's knowledge of ultimate reality; the ideas are always derived from other sources. Thus the experiences add nothing new and at best merely confirm the previous beliefs in a circular manner (since the interpretation of the mystical experience derived from the same interpretive scheme). This position overlooks the fact that the initial beliefs developed in light of experiences, with the mystical ones being deemed of primary importance. They provide an awareness of the world most often deemed more fundamental than empirical knowledge if they are taken seriously. Because there is an interplay of experience and interpretation throughout the development of any conceptual scheme, mystical experiences (or any other experience) are not a totally independent or fresh source of knowledge, but they remain a unique mode of experience that expands one's view of the world. In addition, when a new experience does occur in a culture whose fundamental interpretive scheme did not evolve with mystical experiences, mystics often have to reinterpret many basic claims as "symbolic." Thus Śaṁkara's interpretations of Vedic creation stories and many Upaniṣadic passages are forced, since the texts are not uniform nor always nondualistic in intent.

During the deeper type of mystical experience, mystics abandon totally their own presuppositions and interpretive schemes—if the experience is truly void of all conceptual content as mystics claim—thereby permitting the presence of the "wholly other" mystical. Thus depth-mystical experiences may be the one allegedly cognitive experience that fits the empiricist model of experience.[8] Interpretations, i.e., structures of understanding imposed either through reflection or unconsciously, arise only when discrimination is again present. Thus the mystical is directly experienced, not inferred, but understanding is not present during the experience. After the experience, some interpretation is necessary in order to understand the experience and its significance. What is taken to be the *insight* combines elements from the experience and from the conceptual scheme. Mystics normally equate the conceptualization with the experiential component, that is, they confuse the over-whelming sense of importance that the experience gives with the strength of the interpretive element. Since the experience is given embedded in a conceptual structure, the sense of "absolute certainty" or "self-validation" easily becomes misapplied to the interpretation. In science, the roughly corresponding danger is of ideas becoming *a priori.* The interpretive schemes do not appear to their users to contain insight-claims that need to be defended. From inside a belief-system, it looks as if we are merely describing, not interpreting, the given.

Mystical systems run the risk of being self-fulfilling, that is, setting up a view of reality and confirming themselves by means of this view. All metaphysical systems (e.g., naturalism) run this same risk. This self-certifying quality may be present in all conceptual schemes to a degree. For example, Clifford Geertz sees religious systems as defining a reality that they use in turn to justify themselves.[9] Feyerabend says the same about science: empirical "evidence" is created by a procedure that quotes as its justification the very same evidence it has produced; he likens empirical theories to the claims of witchcraft and demonic possession developed by Roman Catholic theologians in the fifteenth to seventeenth centuries in Europe.[10] Once enlightened, we no longer appear to have a "view" but to see things as they really are. There is an "aura of factuality" about how *we* see the world. All experiences "verify" the beliefs from inside a circle of faith much as Ptolemaic astronomy was verified by every predicted eclipse before a plausible alternative interpretation was advanced. The more metaphysical the system of thought, the less open are its adherents to a genuine consideration of alternative systems.

The only countermeasure to this self-confirming process is to realize that all construals of mystical experiences are open to question and none is absolutely confirmed by the experiences themselves. All experiences fall into the same situation—none carries with it its own interpretation. For instance, even if it is argued that "self"-awareness (awareness of one's own immediate state of awareness) is the one certain cornerstone of knowledge that we all have, still it is open to interpretation: Descartes takes it as evidence of an abiding mental entity; the Buddha takes each act of consciousness to be separate and takes the notion of an enduring self as an unverified posit; and Śaṁkara takes this background awareness to be the reality underlying all "objective" reality. Nothing about mystical experiences places them in a privileged epistemological position. Even assuming there is only one basic depth-mystical experience, the fact that there are varying conceptual schemes within which this experience has been incorporated would indicate that the experiential element in mysticism does not provide the structuring in an absolute and straight-forward manner. (Nature-mysticism involves both different experiences and different schemes.) The conceptual element necessary for understanding our experience comes from outside any one given experience and develops in conjunction with it. Thus knowledge even in the mystical realm is a human product, with contributions from the knower and the known. "Facts" as the content of knowledge are produced from an interaction of the knower and reality here as in science. This goes against Śaṁkara's claim that ordinary and Brahman-knowledge depend exclusively upon the object of knowledge, and not at all upon the knower (BSB I.1.2,4). Instead, in light of the plurality of mystical systems, it seems more likely that mystical experiences, like sense-experiences, are "signs" whose meaning must be interpreted.

Thus we must agree with Smart that the identification of the self *(ātman)* with Brahman comes not from inspecting the experience itself.[11] The self may, in fact, be experienced, not inferred, but its status—its relation to "objective" reality—depends upon other considerations. Similarly, the branding of ordinary experiences as "illusions" in the pejorative sense or as not being insights into the nature of the world also reflects nonexperiential judgments, even if the claim does appear to be given in the experience itself. Such claims cannot be deduced from descriptive claims alone; experiences may motivate a decision to accept or reject a claim but, as Popper says, they cannot prove or disprove it.

Staal has an illuminating discussion connected to this issue.[12] Awareness of Brahman *(brahma-vidyā)* and ordinary consciousness are incompatible, but all philosophizing occurs during the waking state. Therefore, preferring the Advaitic experience is itself an act of ordinary consciousness; all our knowledge and interpretation of this experience occurs in the waking state only. So also the doctrine of the illusoriness of ordinary experience is the outcome of speculation in this state of awareness, since no doctrine can come out of the mystical experience itself. Speculation can lead to a consistent philosophical doctrine, but cannot establish its own truth. Experience alone is decisive for becoming *convinced* of its truth. But experience is not a decisive proof in a logical sense: the interpretation Advaita gives to the mystical experience is consistent with the tenets of the system, but neither is based upon the other in a philosophical sense. Staal further notes that Śaṁkara never invokes any mystical experiences as proof of his doctrine—awareness of Brahman is given in an immediate experience *(anubhava),* not by one of the means of correct knowledge *(pramāṇas),* which operate only in the realm of the dream.

The Vedas are the final court of appeal for Śaṁkara with regard to the mystical. We are aware of Brahman through self-awareness, but the Vedas provide the proper understanding. Because Brahman is not an object of the senses, the other means of correct knowledge cannot obtain (BSB I.1.2). Thus, closer examination of the world will not validate the interpretation. By this appeal to scripture, Śaṁkara becomes one mystic who admits that experience alone is not enough to justify one's position. This may seem to be dogmatic, but experience alone cannot decide all issues, even in science. And another problem arises: Śaṁkara takes the Vedas to be revealed to the ancient seers, but even revealed scripture needs interpreting. For example, Śaṁkara resolves literal contradictions between the Vedas' creation stories by differentiating the primary from secondary meanings (BSB I.1.14). This necessity to understand reintroduces the basic problem of the relation of concepts to experience.[13]

Smart feels that the *truth* of a doctrine depends on evidence other than the mystical experience.[14] But probably a better way to view the justificatory process is to view it as the same as the discovery-process: elements from experiences and the concern for understanding interact in developing a conceptual apparatus that adequately accounts for our experiences in light of our concerns. Having "reason to believe" an interpretive system will depend upon both experiential and conceptual considerations.

There can be no experiments to decide between alternative mystical systems because the mystical experiential contribution remains constant, that is, new experiences merely would repeat the past contribution. So too, mystical experiences seem equally open to various theistic, nontheistic, nondual, and pluralistic options of interpretation. On the conceptual side of the problem, because the frameworks within which mystical experiences are fitted are the broadest court of appeal for any argument, disputes as to which system is the "correct" interpretation will remain unresolved. Debates between mystics on doctrinal points reflect more a conflict of visions of reality than an open consideration of alternative views. Certainly there is no progressive "research program" such as Imre Lakatos ascribes to science.[15] There is no series of theories, with each later theory having a greater corroborated content and explanatory power than its predecessors. Nor are mystical and nonmystical alternatives falsified by comparison with each other. The central "ideals of natural order" within a mystical system need not be accepted uncritically—alternatives and possible objections are discussed by the Buddha and Śaṁkara. But these ideals will be the last court of appeal, determining what is accepted as a fact and a reason. There does not appear to be a standard independent of specific ways of looking that can be invoked to settle the issues. For such a settlement there would have to be some theory-neutral way of determining the way things "really are." But determining this involves reference to the workings behind appearances, and of these there is no theory-independent account or method to resolve conflicts on a metaphysical level comparable even to the imperfect means of theory-acceptance in science.

Because no world-view is capable of final proof, a normative judgment becomes involved. Vindication of a whole value system may be possible only by appeal to a way of life as a whole to which we are committed. All that can be done to justify anyone's adopting a value system as a whole may be to invite others to adopt it, that is, to see whether these ideals reflect what they want to see realized in the world and how they want to live.[16] Nothing comparable occurs with science.

Kamma and Rebirth in Theravāda Buddhism

The alternative position is, to use David Kalupahana's characterization, to see Buddhism (or any other mystical system) as antimetaphysical and empiricist, since it does not accept anything that cannot be experienced through the senses or through paranormal powers.[17] The traditional way of saying this is that the Buddha had no theories, no metaphysics, because he saw the world as it is. K. N. Jayatilleke's whole book, *Early Buddhist Theory of Knowledge,* is an attempt to show that the Buddha was a positivist and verificationist, concerned with immediate experience and ignoring metaphysics as meaningless.[18] The Buddha advanced only verifiable hypotheses, and had a "trial and error" experimental method; he differed from his Western counterparts only in that he admitted paranormal experiences as cognitive.[19]

Jayatilleke and his student Kalupahana are asserting more than simply that the Buddha was interested in what can be directly experienced. The former portrays the Buddha as a logical empiricist, and the latter as an empiricist. Kalupahana uses "empiricism" to refer to the Buddhist claim that the six senses (the mind being the sixth) and their objects are everything.[20] But this is itself more a metaphysical position than the empiricist theory of knowledge. Classical empiricism contrasted with classical rationalism in maintaining the sense-experience is ultimately the only source of knowledge in *a posteriori* matters. So too the mind was seen as a *tabula rasa* (contra Buddhism on nescience and dispositions). In the current philosophy of science, empiricism has an established usage: that theory-free data determine the selection of theories, et cetera. If all that the Buddha had was interest in experiences rather than speculation as central, then "empiricism" is too loaded a term to apply.

To see if this is true, let us shift from the issue of justifying an entire mystical system to justifying only a few elements of one—*kamma,* rebirth, and dependent origination in Theravāda Buddhism. If these doctrines do not fulfill the empiricist ideal, then it is safe to conclude that Theravāda Buddhism as a whole does not. Nor, if so developed a tradition fails on these detailed central points, should we be hopeful that any mystical tradition fulfills this ideal.

Can Buddhism firmly justify its claims concerning rebirth solely upon the grounds of those experiences common to all who have undertaken to have them? Jayatilleke's general contention is that *kamma* and rebirth can be experientially tested and even proved through two of the higher knowledges *(abhiññā)* attainable through a concentrated mind: retrocognition (seeing our own past rebirths) and knowledge of rebirths for anyone in accordance with one's own deeds (seeing the law of *kamma*).

But problems arise. Consider Jayatilleke's claims that all the various notions of a self arise from limited and misinterpreted meditative experiences,[21] and that the Buddha considered it possible to misinterpret extrasensory knowledge by drawing erroneous inferences from it.[22] An example of this error is that of those who see conduct in the lucid trances *(jhānā)* but make the wrong correlations (e.g., bad conduct with good rebirths) and generalize from a few instances. Jayatilleke says that Buddha explains exceptions by good and bad deeds from lives before the immediate past life. Kalupahana feels *prejudice* causes some Indians who had knowledge of past lives not to see the proper connections, that is, other ascetics had the same experiences but reached diametrically opposed conclusions (i.e., deeds do not affect our future existences) by ignoring *possible* factors that *may* have been present and that the Buddha considered important.[23] Introducing the idea of possible factors raises the problem of how we could in principle test such claims. How do we know that these possible factors did in fact obtain? And, more important, if others made errors, then the knowledge obtained in these experiences is not self-evident. How do we know that the Buddha was free from error rather than another meditator? How do we test? If the chain of rebirths is beginningless,[24] then how does the Buddha know he saw all of them? And if there are an infinite number of rebirths, then however many the Buddha saw will be statistically insignificant compared to infinity; as in science, no induction in such a case will be guaranteed. And by the Buddha's method of explaining the exceptions (i.e., we need only to look farther back to find the right factor), we could explain any correlations we like: we need only look far enough back in the chain and give weight to whatever we consider important. It begins to look like a theory-directed idea (if not downright arbitrary) when no obvious or immediate connection is observable. We could "verify" not only the Buddhist theory of connection between rebirths, but any connecting mechanism we like by selecting the possible factors appropriately. Since all disputants supposedly have had the same experiences, the experiences themselves cannot decide the issue. The prescriptive metaphysical nature of Buddhism is again apparent.

Even the idea of *rebirth* is not self-evident to all those who have had exactly these experiences: materialists, as Jayatilleke says,[25] who have attained the concentration-states free of perceiving form *(arūpajhānā)*, which gave what the Budda took to be evidence of rebirth, regard these experiences as illusory in the sense of nonveridical or as private experiences giving no knowledg of nonsubjective states. Furthermore, did the Buddha see individual *chains* of rebirths? If so, this would seem *prima facie* evidence for a self with various incarnations—the no-self doctrine would appear more forced and prescriptive. And if we do gain knowledge of past lives, how do we know if it is of *our own* past lives? It could be more simply understood as just paranormal knowledge of other people's lives. Kalupahana himself cites Ayer as saying that we should *prefer* to say that one person picked up a dead man's memories and dispositions (which are still present and open to experience now, somehow) than that he was the same person in another body.[26] Thus hypotheses in this area cannot be decided by experiences alone. There are scientists who accept certain empirical evidence as supporting *prima facie* the hypothesis of rebirth.[27] But arguments can be presented on both sides of the case. To mention another type of experience in this area: near-death experiences are interpreted as evidence of a self existing apart from the body and of life after death, while others studying exactly the same cases with the same thoroughness see them as the products of the brain undergoing its death throes. It is difficult to think of an experiment that would support one hypothesis over the other. And the matter of rebirth may remain open with regard to *all* experiences open to us.

In light of this, it is difficult to maintain, as Jayatilleke and Kalupahana do, that rebirth is directly verified by the Buddha in the simple, straightforward manner maintained by logical positivists.[28] The problem is not merely one of the unattainability of absolute certainty for any one position. Rather there is less consensus here than in more limited scientific questions because the issues impinge directly upon our fundamental beliefs about the construction of reality and of a person. Metaphysical positions are in direct conflict on the issues. Thus complicated arguments would have to be advanced that would involve evidence other than the disputed experiences to support our positions.

Another question of central importance to the Buddhists' religious quest and thus their way of life is whether the Budda can confidently claim to know that his cycle of rebirths is ended. He claims not only to have known directly the end of craving, but also to have known and seen *(ñāṇa-dassana)* that his salvation was unshakable, that that was his last birth and that there would be no further birth (M I.167, III.162). Assuming the previously questioned belief that he verified in the trances that there is a process of rebirths, how does he know that the process has ended for himself? Is it a matter of self-evident experience or of inference from metaphysical principles?

The basic argument advanced in favor of the former option revolves around the premise "Craving is a necessary condition for rebirth." If craving is necessary, then by eliminating it (and thereby becoming kammically natural), the process of rebirth is destroyed. Assuming we can isolate and experience craving, it is reasonable to assume that we can experience the total absence of craving.[29] But if this is all that is experienced in enlightenment, then freedom from rebirth is an inference involving the premise stated above: craving may not be a *necessary* condition for rebirth. The end of rebirth is therefore not explicitly experienced and remains a tentative belief, a claim dictated by *theory* rather than experience. If craving turns out not to be a necessary condition, then the Buddha could still be stuck in the cycle of becoming. That is, even if by surveying his past lives, the Buddha found craving to be the condition for his rebirths, it may turn out to be only a sufficient condition (if it is truly a condition at all) and other conditions will perpetuate the cycle. The conviction and certainty the Buddha had from enlightenment may be another case of an insider taking his view of the world to be the way the world really is. He would not see his implicit inference *as* an inference but as given along with the removal of what he sees as the cause of rebirth. Mystics commonly speak of becoming free from craving, but it is only mystics who operate with a previous belief-commitment to rebirth who speak of an end to a cycle of rebirths.[30]

Therefore Smart again seems correct when he says the claim that *nibbāna* involves the cutting off of rebirth cannot simply be defended by reference to meditative experience.[31] We are inferring from an experience a future change in the ontological status of the "experiencer"—how we see the world is not merely transformed (i.e., an epistemological change), but we have got out of the realm of suffering, death, and rebirth. For this, we would need to show that craving is necessary and that rebirth is at least a plausible theory—something not given in experience simply. Thus the Theravādins do not fulfill the ideals of either empiricism or naive realism with regard to *nibbāna*, even if there is no discussion of any possible positive aspects to *nibbāna* after the death of the enlightened.

Focusing attention upon the Buddhist theory of *kamma*, which is involved in the above argument about rebirth, leads to the same result. Jayatilleke feels that *kamma* is derived as an inductive inference on the basis of extrasensory perception and thus is a hypothesis scientifically verified.[32] But this can be questioned. Although instances could be advanced illustrating *kamma*, it would be difficult to test it experimentally. The connection of *kamma* to rebirth brings in the problems just discussed and vitiates its simple predictive potential: deeds may produce fruit in the distant future, and present fruit may have resulted from distant events. No simple correlation within a short time span is necessary. A further complication is that all volition (through thought, word, and deed) has kammic fruit: any of many acts and thoughts could be used to establish any desired correlation.

The other aspect of Jayatilleke's claim—that this is an instance of inductive inference—is also doubtful. As currently understood, inductions from a limited number of experiences play a very limited role in the formulation and justification of scientific laws, let alone theories. But more important, the Buddha did not claim that knowledge of *kamma* is based upon inference. The claim is that he *saw* his own rebirths and *saw* the law of *kamma* operating there and in other chains of births (D II.20). Invoking induction may be an attempt on Jayatilleke's part to make Buddhism sound more scientific or to bolster a weak point (albeit with a now problematic idea), since all laws assert more than what any finite amount of experience entails. Jayatilleke does note that there are inferences in the *Nikāyas,* inferences being recognized as a means of correct knowledge within certain realms. But the example he cites is not related to seeing *kamma* at work, or rebirths, or anything about actually knowing about the *dhammā,* but only that those who "see and know" will do so in the same way the Buddha does.[33] Nor is such inference inductive.

This type of seeing is theory-laden: the Buddha would see the succession of dying and rebirth in terms of *kamma* the way the enlightened Newton would see falling apples as the effect of gravity. We do not see the linking in the way we see objects. The Buddha's beliefs would shape the correlations of deaths and births. In isolated trance-experiences, we perceive a succession of events, supplying the connecting mechanisms (here, streams of rebirth and kammic efficiency) as with causation. The legendary account of the Buddha's life probably reflects accurately this point: the Buddha had a belief in rebirth before he undertook his religious quest, and he had the belief in *karma* common throughout his area of India. He took these beliefs as ordering-principles into the trances; the beliefs did not come from enlightenment or the trance-experiences. These experiences may have fed back upon these initial beliefs, thereby adjusting them, but to speak of the experiences as verifying them in the manner in which empiricists think scientific hypotheses are verified is simply wrong. The comparative little in early Buddhist texts explicitly discussing *kamma* and rebirth supports the idea that they were presupposed, if not self-evident. What discussions do occur attempt to show how the Theravāda notions fit into their system and differ from other similar Indian notions.

To take a more specialized explanatory concept from the Theravāda tradition, consider dependent origination. This is not an "empirical theory of causality" as Kalupahana argues, since it is not causal (as was discussed in the last chapter). Is it "empirical" in that it speaks "only of observable causes without any metaphysical pre-supposition of any substratum behind them"?[34] Jayatilleke sees it as an explanation in terms of verifiable phenomenal factors replacing the self, an unverifiable entity.[35] Dependent origination certainly is a theory, an explanation delving into the workings behind phenomena in order to explain suffering. But even ignoring the questions of how dependent origination could be tested and of whether any theory is verifiable by observation alone, we still must accept Buddhist claims about the power of concentration for making nescience and motivations (dispositions) observable; and the problems of rebirth and *kamma* recur here, since they are part of the explanation. And if the links between the twelve steps are given in experience, still there is no more reason here than with *kamma* and rebirth to think that the links are not given in experience in a theory-laden manner.

The only defensible position is that of Edwin Burtt: the Buddha realized that the starting point of a philosophy cannot be a purely passive principle, as in Western empiricism, but an active interpretation of experience; speculative thought as such is not condemned.[36] Since science as currently understood does not fit the empiricist ideal, it should not be too surprising that a religious tradition does not either. Jayatilleke and Kalupahana portray Buddhism as more "scientific" than science itself.[37]

If mysticism is not empiricist, are these parts of Theravāda Buddhism *scientific* in nature under the contemporary philosophical portrayals of science? *Kamma* and rebirth are mechanisms invoked to explain regularities seen within surface-appearances, even if the Buddhist concern to explain suffering is not one of Western science. But, as discussed in the last chapter, their problems render them more metaphysical in nature than scientific concepts. If these concepts are labeled "scientific," they fall on the speculative end of the continuum of such concepts. In the end, the Buddhist doctrines are "verifiable" only in the same sense that all metaphysical visions of experience are: the vision itself sets up a framework that determines in advance what will be the "objective" facts and what will count as verification.

Mystery

Another topic arising from looking at knowledge in mysticism and science is what, within a system of understanding, is unknowable or unexplainable in principle—that is, the mysterious. The fundamental metaphysical mystery is why *anything* at all exists. As Wittgenstein says, it is not *how* the world is that is mysterious but rather simply *that* the world is. The joy evoked by knowledge and discovery is confronted by this more imposing reality. A more limited sense of wonder may be touched off by even quite ordinary everyday occurrences—the growth of a flower or whatever. Einstein and many others see this latter wonder as the beginning and end of scientific curiosity. But what remains mysterious in science is not what lies at the edges of research, but what seems to be impenetrable by all our explanation, what we have not begun to understand. This may very well involve the familiar and law-governed.

Science is sometimes seen as the enemy of mystery. In Bacon's words, science is a war on the unknown. Its end is understanding and control, and whatever is understood and controlled cannot by definition be mysterious. Scientists break the world up into manageable parts, assume an orderliness to events, and are satisfied when repetitions are found that can be understood (i.e., are expected) in terms of our posits. The complex flux of what is experienced is thereby "comprehended" systematically and simply.

Quantum physics, according to some thinkers, introduces mystery into the heart of science.[38] Indeterminacy means uncontrollability, and every new discovery just leads to more complexity on another level: no layer of matter describable in a few simple laws has yet been forthcoming. Nor is there any prospect that the situation will change. Thus, it is argued, the more we learn in science, the more mysterious reality appears—mystery reappears, only on a level requiring much more study than even before. But there is a difference between frustration and a sense of wonder and awe. Scientists may be fascinated by the intricacies of nature here, but this only leads to further research. Nothing here is beyond scientific comprehension in principle, nor is research stopping. Niels Bohr reflects the scientific spirit in saying that all reality is essentially fathomable, ultimately knowable. The scientifically minded will attack each succeeding problem in this area until it is understood in a scientific manner. Faith in a solution is not abandoned.

This would place quantum physics with all science as a force removing mystery by rendering the whole world homogeneous and comprehensible. The basic scientific attitude is that the world is a puzzle to be solved, however impressive the world may be. Naturalistic metaphysics goes beyond this reducing reality to what is knowable by science, thereby rendering the world, its opponents say, bland. The only mystery would result from the alienation of consciousness from a world of objects. The opposite response, one of wonder, is also extrascientific and may also reflect a metaphysical position (that reality is more complex than any collection of scientific systems could accommodate).

These responses relate to whether reality is ultimately lawful (or deterministic) or not. But even if scientific inquiry could exhaust the mystery of the world in this sense, still the primary sense of mystery (closer to the matter of the that-ness of the world) may involve the lawful reality. Why is nature put together so that simplicity is a criterion leading to acceptable theories? Why is the basic structure what it is? Even why science and mathematics "work" at all is at present itself a puzzle. Or as Einstein put it, "the eternal mystery of the world is its comprehensibility."

Much that is familiar and lawful is beyond comprehension today, although in the future scientists may probe each area. Many things of a basic nature connected with biology and medicine remain to be understood. The metaphysical mind-body problem is an especially fundamental instance: how does a thought "produce" the raising of an arm? Almost everything basic connected to consciousness, perception, and a sense of "I" remains as mysterious for us as it did for the ancient Greeks.

Most scientists today tend to deal less with grand, metaphysics-impinging questions. Rather our science is becoming more and more specialized. Scientists can detect events lasting only a ten-billionth of a second, but many are impressed with the fact that our knowledge, scientific and everyday, is, to use an analogy popular since William James, only a drop in a sea of ignorance. A recent incarnation of the saying is from the editors of *The Encyclopedia of Ignorance:* "Compared to the pond of knowledge, our ignorance remains atlantic. Indeed the horizon of the unknown recedes as we approach."[39] Even the impressive precision of mathematics does not necessarily reveal a depth to knowledge. From Russell: "Physics is mathematical not because we know so much about the physical world, but because we know so little; it is only its mathematical properties that we can discover."[40]

Being struck with awe by scientific knowledge is one reaction. The opposite reaction relates to the lawful and insecurity. Scientists often find new data that show we do not accurately understand what was thought to be known. This, one would think, would keep us open-minded—we would keep our options open for ways to investigate a largely unknown world. But there seems to be a psychological need to have at least the impression of understanding, which science satisfies for scientist and layman alike. Mach speaks of natural laws as being a consequence of our psychological need to find our way in nature and to avoid having to confront it as confused strangers.[41] But more strongly than this, the metaphysics of naturalism may result from an uneasiness about the unknown. Minimal correlations of consciousness and brain functions, and predictability or other order in all matters are enough comprehension. What we do not know or understand can then be subordinated to the known. Only indeterminacy presents a problem. Abraham Maslow thinks science can be a means to psychological growth or the most nearly perfect defense against facing the unknown yet devised.[42] Science has also been likened to counting the commas in the book of nature, rather than to reading its message;[43] we thereby alienate ourselves from nature, and lose ourselves in a quest for more and more knowledge about smaller and smaller areas of isolated specialization. We let the simple classification of entities substitute for the full complexity of reality. Ours is a Rumpelstiltskinian situation: by knowing something's *name* we control it. Labeling does lead research, but it does not give understanding. Still we feel we have a handle upon a mystery if we give it a name—especially a Greek- or Latin-derived name.[44]

In this matter, Erwin Goodenough took an extremely pessimistic view: we really know nothing basic and are helpless before the forces of nature; scientific accounts are curtains dropped between us and the mystery of reality—we accept the latest scientific accounts credulously because to live aware of our ignorance would crush the vast majority of people.[45] A more moderate position is that we may know nothing basic about reality, but science does appear to give us some genuine understanding. Science is a human construct—perhaps a leaky boat on an ocean of mystery, but one that keeps us afloat. Still, all of these possible psychological motives and responses are not part of science as a way of knowing. Science as an enterprise cannot be cited as a way of knowing. Science as an enterprise cannot be cited in support or repudiation of such conclusions of scientists and laymen.

Mysticism has a different relation to mystery. It remains closer to the metaphysical source of wonder—that there is something unexplainable about simply what is real.[46] In science anomalies (phenomena that conflict with our expectations or concepts) give a sense of reality, of something we cannot get around. Thus they may touch off wonder. So too the comprehension of science may be a source of wonder. But this wonder is more likely to be related to the how-ness of things than to the that-ness of reality. In mysticism this sense is even greater, with the awareness of a reality transcending all our conceptualizations—a power "behind" our conceptual world that is ungraspable or distorted by our understanding. For those to whom science has rendered the world mundane, mystical experiences may in fact be the only source of a feeling of the sacred. This need not evoke the awe and wonder of an encounter with the numinous, but an awareness of something "beyond comprehension" is integral to mysticism.[47]

It could be argued that mysticism, like science, tends to end mystery, since it goes beyond the that-ness of reality into the how-ness, too. By its labels, its conceptual schemes, and its analogies to the familiar, mysticism, too, tries to remove the disturbing nature of the "totally other" and thus render it innocuous. But that there is more to the mystical than can be comprehended by our understanding—that ultimate reality is unfathomable—is essential to all mystical systems. Nature-mysticism has the problems of order and the relation of the measurable to the totality of being. The real is not controllable—paranormal powers for the phenomenal are allegedly possible, but even the enlightenment-experience cannot be forced by meditation or any other preparation. In the enlightened state, control is not the rule; rather, contemplation of the flux of phenomena, perhaps with an accompanying sense of particular awe, is. In depth-mysticism the mystical is construed as experienceable but wholly other than all reality captured by concepts: conceptual understanding introduces a mode of awareness foreign to the mystical itself. The experience is just a "shaft of light" from a mysterious "darkness." The mystical depth remains unfathomable, and how the changeless and timeless reality produces the temporal realm and its lawful structures remains a mystery.

6
Experiences

The next topic falls in line after reality and knowledge of reality: the experiences by which we know reality, and subsidiary experiences of importance. However, it should be remembered that mysticism is no more about *experiences*—as contrasted with what is experienced—than is science. Mystics most usually discuss ultimate reality, the self, God, and so on, *not* experiences (their own or in general).

Defining "experience" is as difficult as defining "awareness," "sensing," or "consciousness." And since science, nature-mysticism, and depth-mysticism differ as to the experiences that are to be taken as cognitively most central, defining "experience" becomes a potentially question-begging affair. The depth-mystical experience contrasts with all other purportedly cognitive experiences: it is devoid of all conceptual and sensory content but is still a state of awareness. The dualism of subject and object in science and everyday life can be referred to as "dualistic" experience (i.e., an indirect indication to a subject of a posited reality) to contrast with the mystical types of experiences. Even the most extraordinary scientific observation (e.g., in the submicroscopic realm where the observing process becomes part of the event) does not fall outside this category. Nature-mystical experiences stand between ordinary and depth-mystical experiences: they are still sensory, or involve the awareness of a mental image, but are not so strongly guided by concepts. Earlier chapters discussed the general nature of each category of experience in more detail. In short, to modify Martin Buber's terminology, if nature-mystical experiences can be called "I-Thou" experiences, then ordinary and scientific experiences remain I-It experiences, and the depth-mystical experience is "I" only (however it is interpreted).

States of Consciousness

Both science and mysticism have a variety of experiences incorporated into their enterprises: science has insight-experiences and observations: mysticism has insight-experiences, sensory experiences, and assorted contemplative experiences. How are these experiences related to various possible "states of consciousness" (different overall structurings of consciousness organized for different purposes)? An altered state of consciousness is not merely a change of knowledge or conceptual understanding within one functional pattern of mental activity, but is an alteration of the pattern itself. Dreaming and the state between waking and sleeping are the most common altered states of consciousness. There may be many possible states of our consciousness, each dealing with reality in a different manner. Huxley made famous C. D. Broad's speculation on the brain as a reducing valve permitting only the data necessary for survival into consciousness—various means may loosen the valve permitting a wider range of data and organizations.[1] Famous too is James's comment that our normal "rational" consciousness is but one special type of consciousness separated only by the flimsiest of screens from other forms.[2]

The idea of "states" of consciousness is of course not new. For the *Upaniṣads,* the self is not attached to any particular state of consciousness (CU VIII. 7–12), but is that power whereby one perceives the waking and sleeping states (*Kaṭha Upaniṣad* IV.4). The *Māṇḍūkya* differentiates four states *(sthānas)* of consciousness: waking, dreaming, deep sleep, and nondual awareness (*tūriya,* "the fourth") which is also the ground of the other three. We in the West do not consider dreamless sleep a state of *consciousness,* but these Indians consider consciousness to be still present until death even if we are not using it. The fourth state is a pure form of the consciousness present in the other three: concentration of consciousness is necessary in each state since there could be no focusing of attention without it. Consciousness is realizable as a fourth state when it is concentrated without structuring concepts or sensory content.

Applying the idea of states of consciousness to the nature of science and mysticism, we see that science operates primarily in the ordinary state of awareness, the state geared toward physical survival. Dream-states and other altered states may be the source out of which some scientific insights spring, but insights must be formulated into scientific hypotheses and be tested in the ordinary state of awareness. Some such altered states of consciousness may even be those entered by mystics, but the ordinary state remains central in science. Similarly, concentration is essential to doing science—Newton and Einstein may have been great scientists because of their ability to focus their attention for long periods of time upon limited problems—but such concentration upon problems never leads to those altered states of "one-pointedness" with or without an object. A change in theories may entail new experiences (a Gestalt-like switch occurring), but these experiences do not disturb the fundamental state of consciousness: the new experiences are different events in one longer-lasting medium (the ordinary state of consciousness). Our science is a product of this one state of consciousness: it produces a picture of the world in which our consciousness does not affect reality as a causal factor. To that extent, science presupposes a state of consciousness as a prescriptive stance with regard to what is considered cognitive. No reference to a state of consciousness being required for scientific research is necessary, since it is not a matter of dispute—it is always only the ordinary state. Charles Tart speaks of the possibility of altered-states-of-consciousness sciences.[3] The central problem with this idea is whether one can keep the theory-directed attention necessary for scientific testing without destroying the experience occurring in the altered state of consciousness. For any scientific observation, a subject-object differentiation may be necessary; but this may not be maintainable in altered states of consciousness. So too any attempt at manipulation may destroy these altered states.

With regard to mysticism, it does make sense to speak of mystical experiences occurring in altered states of awareness. In ordinary awareness, different experiences come and go without always affecting the fundamental organization of mental activity. But with mystical states, the experiences occupy awareness to such an extent that changes in the experiences tend to return the experiencer to normal states of consciousness. A different conceptual element is not simply applied to the same basic everyday awareness; instead, studies of physiological indices would indicate (at least indirectly) a different structuring of mental functioning entirely. For nature-mysticism, it still makes sense to distinguish experiences from the longer-lasting states of consciousness. Enlightenment may involve a permanent change in consciousness, as well as other changes. For the depth-mystical experience, on the other hand, any possible change in consciousness converges with the experience itself—the Advaitic unity would continue. That is, this state of consciousness, unlike others, is identical to the experience and what is experienced. Sometimes mystical experiences in general are referred to as involving an element of the mind "deeper" than the diffused "surface" awareness of dualistic awareness. But the notion of different organizations of awareness for different approaches to reality is at least as illuminating.

Yet discussions of differing states of consciousness are of limited importance to mystics, since ways of life are central to their concerns. A concentrated state of mind is not itself the required insight-experience or knowledge informing a way of life. It may be necessary to, or may facilitate, the experiences connected with the knowledge-claims, but it is not equivalent to them. The possibility of different *experiences* may be important since the *mode* of knowing, not merely the knowledge-claims, is important in mystical wisdom. But, as discussed earlier, experiences alone do not determine the knowledge-claims, and introducing the matter of the modes of consciousness out of which such experiences may arise is another step removed from the concern with knowledge-claims. The *Māṇḍūkya Upaniṣad* discusses such states, but they are a minor topic in the other *Upaniṣads,* and even in this text the relation of the different states to the self is of central importance. For the Theravādins, the interest is in ending suffering, not in gaining interesting altered states of consciousness. Insights occurring in such altered states are seen as applying to the world independent of the state of consciousness of the experiencer. And the insight required for enlightenment may in fact occur in the ordinary state of awareness—the mind is more "concentrated" than normally, but enlightenment can occur only outside the trance-states of concentration. The resulting internalization of a world-view may result in new experiences and even in a new state of consciousness within which individual experiences occur, but it is the end to rebirths that is important for the Theravādins. Switching to the matter of the state of consciousness may reflect better our contemporary Western interest in psychology, but it does not reflect the mystics' interests very well.

Going into the bodily mechanisms connected to consciousness is even farther removed from the mystical interests. The hemispheric specialization of the brain is often mentioned in discussions of science and mysticism: in right-handed people, the left hemisphere specializes in language, temporal divisions, and analysis in general, while the right's activity is nonverbal, spatial, and synthetic. The left is active, the right receptive. Thus science is said to be the product of the extreme instance of the manipulative specialization, and mysticism of the receptive mode of awareness. The attempt to grasp the mystical with the analytic mind is doomed from the beginning because it introduces a fundamentally different mode of awareness: only if the mystical were an object among objects could it be an object of scientific scrutiny. Such a physiological account is at best a scientific explanation of mystical experiences, not a "translation" of mystical ideas into a scientific idiom: if the theory proves groundless or becomes indisputable, still the issue of whether mystical claims are insights constitutes a separate topic of discussion. Specialization of the brain-hemispheres is simply irrelevant to such claims as that the enlightened have escaped the cycle of rebirths. (See the appendix.)

Insights

Altered states of consciousness in science are most often mentioned in the creation of new theories and new approaches to problems. Deduction or induction or any simple reasoning process from observation or from old theories will not produce the radically new visions constituting many new theories. Fresh insights are a matter of imagination; and intuitions may arise from anywhere, with experiences in altered states of consciousness being one source. Creativity often consists of seeing that an unnoticed assumption is only an assumption and is in fact wrong; the initial problem is then rethought. Or creativity may involve connecting phenomena that appear diverse at first glance. Newton, for example, brought together falling objects, tides, and the motion of celestial bodies under one explanatory concept. This concept (gravity) involved a notion (action-at-a-distance) that appeared absurd even to Newton. Many new ideas appear absurd: the idea of creation *ex nihilo* in steady-state cosmology, or the idea of multiple worlds to explain certain submicroscopic phenomena. In fact, if a truly new point of view is being advanced, it must appear odd at first—it must be free to a degree from the conventional scientific ways of thinking of the time. New concepts are hard to understand or even express at first. This makes the process of creation even more remarkable. Theories as presented in final forms express the weight scientists give to observation and testing; the imaginative guesswork is gone. But this should not produce the impression that science is a matter of machinelike computation alone rather than being primarily the work of imagination.

Insights in mysticism differ from those in science in significant ways. Enlightenment-insights in mysticism lead to new ways of living (at least the cognitive and dispositional structures of mystics have been altered), while in science merely new theories evolve. The enlightenment-event is of central importance since it is a climax in the mystical ways of life, while in science only the final theories are of importance. In addition, mystical experiences are more confirmations of previously held beliefs than fresh conjectures or new solutions to old problems that add to our body of knowledge. Śaṁkara, for example, does not claim any revelatory experiences. Nor do these insights become shaped into hypotheses to be confirmed or refuted by other experiences or to be argued about more generally; the mystical approach is more absolutist and metaphysical. The enlightenment-experiences provide the conviction for the future way of life and for the interpretation of the experiences therein.

The insights constituting enlightenment, such as the experiences of the night of the Buddha's enlightenment, do fit a psychological pattern for insight-experiences in general: intense study of the knowledge of the day (be it the current scientific research or the religious knowledge of one's tradition or environment), a period of relaxation (letting the "unconscious" mind work over the data), and a sudden unexpected insight.[4] Concentration on the problem is necessary as a first step followed by release from struggling with the problem, whether in exhaustion, wonder, or just changing the subject of one's attention. The insight itself often seems to come from outside oneself—an "in-spiration" from the gods—although the view that the solution comes from the workings of the mind outside our awareness at the time seems more compatible with our general view of reality. In science the two further steps are the shaping of the insight into a testable hypothesis and the subsequent testing.

In the empiricist philosophy of science, a distinction must be made between the context of discovery and the context of justification: the former may involve intuitions from any source but only the latter (the testing of hypotheses) is of philosophical interest. But problems arise for an absolute distinction. The first is that dreams and experiences in other altered states of consciousness are not scientific hypotheses: they must be worked into hypotheses in the ordinary state of awareness. Thus Kekulé's experience in the state between waking and sleeping of snakes circling with their tails in their mouths was not in itself a scientific conjecture. His *insight* occurred when he saw the applicability of this image to the problem of the molecular structure of benzine. He did not see all his unusual experiences as having scientific import. The factors operating in the insight are precisely the conceptual factors involved in justifying the claim. (It could also be argued that the psychological factors producing the dream were the final scientific considerations operating on an unconscious level of the mind—that is, that his unconscious too is shaped by his knowledge and concerns.)

The fact that we need to internalize the current state of scientific knowledge for an insight to occur has led to doubts as to whether there is a nonrational element in scientific insights at all. Almost all the great discoveries—even those of Newton and Einstein—were duplicated (or nearly so) by other scientists working at the same time: all were dealing with similar problems and thinking along the same lines. "External" factors such as social and political commitments may also have eliminated from consideration certain options. At any point, only a limited number of options, not all logically possible theories, are given consideration. But scientific creativity is irrational or nonrational only if rationality is equated with logical deduction. Even if there is no perfect algorithm for producing successful theories, in studying these discoveries there may be found recurring "patterns of discovery" and therefore a "logic to discovery." C. S. Peirce spoke of "abduction" instead of induction or deduction for these discoveries, and suggested that the mind puts limitations on possible hypotheses, that is, there may be in the structure of the mind composed by nature a biological root to the type of explanations we find. The possibility of programming computers to generate plausible new theories gains credence if some such pattern to new discoveries can be found.[5] The basic objection to this is that such a procedure would be too conservative: devising radically new ideals of natural order requires a greater break with past ways of thinking than can be incorporated into any established procedure.

Objectivity

However insights arise in science, they do not all succeed. We only hear of those which have led to advances. But conjectures in science must be tested to one degree or another. And this brings up the matter of objectivity. Scientific objectivity is not a matter of the origin of ideas but of their justifiability. Justification of a scientific (or mystical) claim involves more than reference to experiences alone, as discussed earlier. But the objectivity of phenomena or experiences plays one necessary, if limited, role.

The backbone of scientific objectivity is the ability of replication by others, that is, instructions for public testing. (In fact, experiments are seldom repeated unless the results are controversial; instead, the findings are accepted and the experiments are refined and extended.) "Intersubjectivity" could be substituted for "objectivity," since it is *experiences* that are replicated. The experiences replicated in science occur in the ordinary state of consciousness and within a subject-object framework. But this may not be necessarily so: if procedures could be outlined for reproducing experiences in other states of consciousness or outside the subject-object framework (the reproduction being testable by agreement of description), other experiences would have the same claim to objectivity.

In this sense, mystical experiences may also be objective. Since subtle mental changes are needed, replicating the required experiences may be more difficult than with scientific observations: more sustained effort may be required (although scientific experiments are often difficult to repeat), but this is a difference in degree, not type. The basic hindrance is that no action can force us to become *desireless:* following the steps of a mystical path cannot guarantee a switch in world-view and dispositions. Some traditions, including the *Upaniṣads,* even speak of "grace." This may destroy any claim to objectivity. Mystical experiences may not be open to all but only to those with certain psychological dispositions: the emotional character of a few persons or perhaps many may preclude them from having such experiences.[6] But mystical experiences are not a capricious gift; through psychological studies, we may be able to predict who would not be prone to having them. In addition, the means to replicate the experiences can be specified for all to follow, and if the required conditions of selflessness and so on are fulfilled, the experience does occur in a lawful manner. Such experiences have been induced in mystical traditions for centuries by means of instructions, aids, techniques, and explanations. Tests have been devised by masters to see whether the disciples have truly attained enlightenment. Physiological studies of the correlates of mental events suggest a change in the state of consciousness, and thus may provide objective support for the claim that the experience occurred; but there are philosophical problems as to whether this is conclusive (as is discussed briefly in the appendix).

Other aspects of objectivity seem to be shared by scientific and mystical experiences. First, *all* experiences are private—none can be presented for public examination. For both science and mysticism, we assume that the relevant inner experiences are basically the same for everyone because of our common biological structure and the similarities of descriptions among those persons who allegedly have had the same experiences. Second, if there must be a community sharing a common set of assumptions and concepts in order to reproduce data, as is often claimed now for scientific objectivity[7]—that is, only a community of trained experts, not the general public, can do the necessary observing—then mysticism certainly is no worse off than science. Each enterprise may require many years of training in order to see exactly what is relevant to the claim in question. The interpretation given an experience in both enterprises may be theory-laden, but this does not affect the claim to objectivity in one case more than the other.

A corollary of this concerns the esoteric nature of each enterprise. Not all mystical traditions consider their knowledge "esoteric" (i.e, kept only for the initiates). The Theravāda and Advaita, for example, do not. In addition, the mysticism of all traditions is becoming more available to the public in this century—and science is becoming increasingly inaccessible. Before this century, every intelligent layperson could follow the theories in every field of science without long years of initiation into the sacred language (mathematics) and of training in the mysteries. Mystical traditions in the past may have been more strenuous in keeping the knowledge from the public, but this is primarily because of the belief in the *danger* of the knowledge—even the very words involved possess a *power*. It was simply a matter of keeping dangerous knowledge "top secret," as with the dangerous aspects of science today. It is the more practical elements of science that we protect, not the basic research; but mystics consider all of their knowledge as directly applicable to our lives. In addition to having to screen out those persons who might exploit the knowledge's power for selfish purposes, the masters of esoteric traditions oversee the preparation of the initiates for the handling of the power—bestowing the various "clearances" for access to the knowledge, as it were—and provide an oral commentary for understanding the basic writings of the tradition. But no more than the scientific knowledge of today is mystical knowledge deliberately hidden from anyone who sincerely wants to learn it and who is willing to undergo the rigorous and extensive training required to attain it.

In the end, a condemnation of mystical experiences as less objective than scientific ones is as much a value-judgment as anything. It is harder to fulfill the requisite conditions, since a mystical experience is not related to an object in the customary sensory way, but it is potentially as open to replication by anyone as the experiences related to scientific theories. Mystical techniques cannot guarantee enlightenment, but not all scientific experiments succeed in duplicating the established results. Each enterprise can explain failure (necessary conditions did not obtain). We may accept science and reject mysticism because the former's claims fit in better with our total view of the scheme of things. Mystical claims do not have this coherence for us, and a more readily accepted interpretation is that mystics misinterpret the significance of their experiences—in reality, mystical experiences have no cognitive import.[8] But if mystical experiences are widespread and if those who have given common descriptions (leaving aside the issue of distinguishing the theoretical interpretation from a description of the experiences themselves), then as Ayer says, the criterion of objectivity is satisfied and psychological reductions are unreasonable.[9]

The chief problem with this type of objectivity is that replication can merely reinforce an erroneous understanding of a phenomenon. What is reproduced in scientific observations—what is tested—is not bare sensation but conceptually structured perceptions. The set of assumptions is tested and any errors in them are replicated as part of the data.[10] For example, every predicted eclipse for over a thousand years confirmed Ptolemy's cosmology. Nor is the validity of a claim established by the mere accumulation of such repeated results. If sensation devoid of any concepts were reproduced, the results would be shorn of any understanding-component. The scientific question is whether the stick in water is actually bent, not whether everyone sees it that way. Thus bare sensations would not decide scientific controversies. Experts decide what the "facts" are. The result is a social decision reflecting the commitments of the professional group, with the possibility of merely reinforcing past prejudices (both extrascientific and theoretical).

Mystical experiences share this problem: replicating an experience void of all sensory and conceptual elements does not test one interpretation or the general cognitive significance of the experience. The significance for mystics of their experiences, as with scientific experiences for scientists, is decided by the society of experts (meditative masters), not the general public. For mystics, no experience will count against the significance of mystical experiences. So too for metaphysicians who take scientific experiences as of central cognitive value: what experiences would be permitted to count against science in principle? The objectivity of the experience alone will not satisfy the claim to reality; other considerations are involved.

Intersubjectivity does not attempt to make us nonhuman with its emphasis upon such types of observations. The hope of this type of objectivity is to remove personal prejudices and other emotional responses not essential to the replication of an experience, but observation still remains a human phenomenon. The role of decision by the community attempts to reveal what is purely a personal reaction and what is common observation (although the problem of group's prejudice replaces that of the individual's). The impersonal ideal, that is, stating claims without reference to a speaker, holds in many clearly human enterprises—for example, in ethics.[11] In mysticism too the claims are usually held to be true for everyone, not just the experiencer.

Of course, what is objective may not be of utmost value in our lives nor the most revelatory of reality. For one thing, the unique (something unrepeatable in some respect deemed important) would be omitted. Thus to take the objective as essential to a cognitive standard is a normative judgment. To live accepting *only* this one abstraction as providing an insight into the nature of the world may also be unsatisfactory. Nevertheless, intersubjective testability does not attempt to provide a solid base of recurring points upon which we can base our lives. This may even be necessary for life, even if it is not the substance of it.

Scientific Observation versus Mystical Contemplation

The general conclusion is that science is not so "objective" and mysticism not so "subjective" as they might appear. This is reinforced by comparing scientific observation with nature-mystical experiences of the world. In fact, in one important regard mystical contemplation may be more objective than scientific observation.

In principle, scientists are *disinterested* in the outcome of experiments and tests. That is, they should have no personal attachment or selfish desires connected to any findings. Attaining such a state of mind in practice is very difficult. It has even been likened by Popper to the mystical purification of the soul. Not only in testing claims, but in the creative conjectures would such disinterest prove of value. But as background beliefs affect intuitions, so also may desires affect scientific results. For example, there is the case where a group of research assistants were each given two groups of rats that were in fact virtually indentical in abilities; but the assistants were told that one group of rats would perform better than their other group—and that is precisely what happened. In addition, the scientific mentality has been described as "a vehement and passionate interest" in the relation of general principles to facts.[12] So too personal reputations and research grants may be at stake in individual research projects. In such circumstances, it may be too idealistic to speak of disinterest or nonattachment to the outcome of one's work. Emotions and desires drive artists, and this is true of scientists too. Science may even become the source of meaning for their lives.

But if scientists do achieve true disinterest during their work, they are not *uninterested,* that is, they may not be selfishly involved, but lack of personal attachment does not mean that they can be totally indifferent to what goes on in their work: there is a minimal need to direct attention to particular problems. Science as an enterprise has particular interests and limitations; attention is further restricted by individual theories. If scientists were not so directed, no insights would ever occur. Thus interest is required even if no emotional profit is involved. Likewise, there is a creative act involved in knowledge, not a passive reading off of what is there. The lack of a simple falsification of theoretical constructs by counter empirical evidence shows that scientists are not indifferent to their concerns but have a degree of tenacity. Judgments of significance are made all along the way, and include even verification.[13] Even scientific observation is not strictly passive; instead, concepts from each scientific scheme direct perceptions, screening stimuli by directing attention primarily to particular aspects.

Experiments in fact *interfere* with phenomena (controlling some variables) in making observations. This reveals an aggressive, not merely active (let alone passive), stance toward reality by scientists.

The ideal that science is interested in phenomena "for their own sake" rather than for human purposes is approached only by theoretical science, and even here science concerns an explanation, an understanding that puts our mind at ease for the time. On science's more practical side, prediction and the resulting reliability has made it possible for people to have more and more control over nature through technology. The scientific stance places the individual over against the rest of the world where what is needed for material well-being can be discerned and obtained. At least this aspect of science makes it essentially "power-thought," as Russell felt.[14] Even without endorsing the extreme view that we are the masters free to manipulate and exploit the world, still a strong practical component has always been present to science, whether it involves determing fixed astronomical phenomena for religious calendrical purposes in the ancient Near East or the very close ties science has today throughout the world to the military. All in all, science is tied to human interests. Scientists view the world with specific questions in mind.

Nature-mystical contemplation of nature has the features of disinterest, and is in fact closer to uninterested observation. Its disinterest is greater in degree than that of science: all desires, all attachment to the fruit of one's action, and any hint of selfishness are to be rooted out completely. This, unlike in science, is an absolutely central objective in mysticism. Removing a sense of self may be useful for science, but the most radical accomplishment of this occurs only in mystical traditions. Even one's own actions are observed as if they were not one's own. Total uninterest is also attempted. Observation is not captured by conceptually highlighted points. The resulting perceptions do not reflect attention to "this" or "that." It is only in the extreme form of nature-mystical experience (i.e., totally conceptually free sensation) that such complete uninterest can be achieved and a mirrorlike observation occur. The mind is totally passive (receptive), accepting whatever is presented without cognitive or emotional judgments, and without expectations or hopes. Such a "pure" nature-mystical experience, unlike the enlightened state, could fulfill the empiricist ideal of sensory experience free of all theory (although awareness would not be directed to particulars).[15]

Complete concept-elimination does not occur in the enlightened nature-mystical state, however. Rather an alternative conceptual scheme to everyday beliefs and a different attitude toward concepts are permanently internalized. This precludes complete uninterest (as was discussed previously concerning Mahātissa's enlightenment). Whether mystics see the world "as it really is" is ultimately as open to question as the case with scientific observation. The concern for salvation is a human concern that may be more related to our emotional states than the human concerns dealt with in science. The role of knowledge in the early *Upaniṣads* in fulfilling desires bridges the gap betwen these religious and any scientific concerns; the later, more mystical *Upaniṣads* play down this aspect. In general, mystical observation is less related to prediction and expectations. Scientific observation occurs in the framework of ordinary, conceptually differentiated perception, and theories add more conceptual differentiations. In principle this is not more "objective" than mystical observation. Even scientific insights occur only after a set of beliefs and problems have been internalized; and these insights must be formulated and tested in the same theory-directed manner. Mystical enlightenment too comes with the internalization of a point of view setting up conceptual screens; but its lessening of conceptual control of perception in general causes it to be more detached in this regard than scientific interest. Thus any contrast between religious commitment and scientific detachment in this regard breaks down.

Mystical objectivity is in fact incompatible with the attention directed by concepts in scientific observation. To achieve this passivity is actually a strenuous feat. The person remains very active in meditative exercises before the state of relaxed receptivity is accomplished. In science one remains active throughout the entire process of observation. Thus the mental "energies" are employed at different places in scientific and mystical observation, just as concentration in science requires conceptual structure while concentration in meditation attempts to eliminate all name and form. At no point can science achieve such a state of surrender and still remain science.

Scientific and mystical observation are on a continuum, with mystical experience tending toward less conceptual control ("participation" of the observer "in" the observed) and scientific experience tending toward total control by concepts (observation as perceiving that a category is being filled). Thus science tends to restrict awareness, and mysticism to open it up. The exact point on the continuum at which mindfulness rules out the attention to "this" and "that" necessary to scientific observation is not clear, but it is clear that one cannot be a scientist and practice the "pure" form of mindfulness (free of all concepts) at the same time. The mystical ideal of uninterest also makes its observations unsuitable for scientific use—it is an "objectivity" that would never lead to scientific discoveries. Bare attention would never make the necessary connections between observed phenomena nor has it the interest in the structure behind appearances necessary for scientific discoveries.[16] In the extreme instance of nature-mystical experiences, the stick appearing bent in the water is the stick "as it really is"—no reference to a "real" structure behind appearances is made. With this type of experience, there is more an attempt to "live with" the world rather than set the world over against the observer as an object to be analyzed, controlled, and explained to our satisfaction. Ironically, this leads to a reversal of the *prima facie* conclusions: mysticism in general observation ends up more objective in the sense of being indifferent to what is there, and science more concerned with human considerations.

7
Language

A very illuminating topic for comparison is the scientific and mystical uses of language.[1] Similarities arise here because each enterprise deals with phenomena occurring outside the range of the everyday world (in science, large- and small-scale empirical phenomena; and in mysticism, nondualistic and concept-free experiences), and each deals with problems of conceptual systems (in science, the replacement of old concepts with new ones; and in mysticism, metaphysics and the status of any conceptual understanding). Mystics are more aware of problems with language because science and language are parts of the point of view separating subjects and objects, while mystical experiences are not. Thus ineffability is often claimed in mysticism, while problems of the changing meanings of fundamental terms are central in the evolution of science. A major difference in their language-uses is that the assertive function of language is central to science; instructions too play a part, but factual claims are ultimately what is of importance. The various mystical factual claims are important but so are the instructions, recommendations for actions, and other evaluations. Often key terms (e.g., "*tao*" and "*dharma*") have both evaluational and factual dimensions. As with the rest of this study, the *assertive* function of some mystical utterances will be the primary focus of attention within mystical discourse, since this provides a common ground for comparison and contrast with scientific utterances.

Meaningfulness of Utterances

Religious utterances are sometimes given noncognitive interpretations, that is, as voicing emotional responses or making recommendations but *not* making claims about the nature of the world. This position is defended usually because religious claims are seen as fundamentally different in nature from scientific claims—and if they are not scientific or checkable in some simple everyday manner, they cannot be assertive. (Even theoretical terms in science have been treated in the instrumentalist position as not asserting what is the case.) Some advocates of mysticism claim that mystical utterances are no more than exclamations. But obviously mystics find language useful in leading others to enlightenment and in verifying that enlightenment has occurred. And their assertive utterances under the least forced interpretation entail factual claims about the world. They are more metaphysical in most instances or related to a different experience of the world, but they are not reducible to recommendations. The risk of a normative judgment is very great in this regard; at a minimum, an argument would need to be advanced for why only scientific assertions can go beyond everyday phenomena. To dismiss mystical claims as meaningless because they do not fit the scientific situation when they do not purport to do so would obviously need a reasoned defense.

Still there is the problem of how to understand the *meaning* of mystical utterances. Even in science the meaning of terms is not unproblematic: theoretical terms or claims are not given meaning in a simple, one-by-one manner, as strict verificationists or falsificationists have contended.[2] Theory gives meaning to experiential claims as much as vice versa. "Simple" observation is itself theory-laden. Theories as a whole give meaning to each utterance: some terms may be more closely related to experiences and others to definitions, but only in the context of a total theory do the parts become meaningful. In this regard mysticism is in the same position as science. The difference is that science makes empirical claims: ordinary experiences are involved in checking some claims within the theory as a whole. Mysticism, however, does not attempt to explain why one set of empirical phenomena occurs rather than another. Hence these claims cannot be checked through normal experience. The only interest in specific empirical phenomena for mystics is as to which course of action will lead to the enlightened point of view. This involves courses of action and the factual claims related to them, but any dualistic experience will still be irrelevant to their factual claims about the mystical and its relation to the world of appearances.

Mystical experiences do appear to possess the objectivity of being repeatable by following a set of procedures, although there are problems here (as discussed in the last chapter). This may satisfy critics who want objectivity for a claim to be deemed meaningful. But experiences do not supply meaning to claims in a simple manner. The fact that mystical experiences are open to widely different interpretations should convince us that the meaning of the claims does not come from the experiences themselves alone. That is, the intention of a claim and how to handle its concepts come from the total conceptual system.

Our standards of what is meaningful for an assertive utterance have been established by our acceptance of science as central. Mysticism does not hold so certain a position in our culture, and so is more problematic for us. But people who have embraced mystical traditions have argued about doctrines and have been led to special experiences. Just because a practice occurs, of course, does not make it legitimate, yet to dismiss mystics' claims because they do not fit the scientific pattern may be ill-advised. And not all scientific claims are directly checkable by dualistic experiences (even ignoring that a theory determines a concept's meaning). It may be that each enterprise must set up its own criterion of what constitutes a meaningful utterance within its context. Making meaningfulness dependent upon the experts of an enterprise raises the possibility of circularity, but any criterion will reflect only certain values and no set of values seems to be universal. Establishing any criterion as necessary for all assertive utterances will run into problems, as with the verificationist and falsificationist criteria for theoretical claims in science alone. Understanding mystical claims may require a certain sympathy, if not certain nonordinary experiences, in order to see what is being interpreted; and accepting the meaningfulness of mystical language prior to understanding may also require a leap of faith. If science dominates our thinking, we may not be able to place ourselves imaginatively within a framework of claims that does not rely upon dualistic experiences in deciding issues.

Mystics, like scientists, do not discuss what separates a meaningful statement from a meaningless one. Some statements are dismissed as false, but none are usually condemned as meaningless. Within the Buddhist tradition, the notion of permanent and independent referents is attacked, not the meaningfulness of the claims. The statements, once we no longer project them as indicative of the ultimate structure of reality, are useful and even necessary in leading the unenlightened to enlightenment. Jayatilleke, in his attempt to portray the Buddha as a positivist, interprets the unanswered questions *(avyākatā)* as unanswered because some deal with matters beyond experiences and some were logically meaningless.[3] But the questions concerning the temporal and spatial extent of the universe are certainly not any more meaningless or less open to experience than other scientific questions. Nevertheless, they do deflect attention from the religious goal of ending suffering and so are unanswered. Any answer to questions related to whether the "self" and the "body" are identical or different and to what happens to the enlightened upon death (e.g., S IV.373f.) would not "fit the case," just as after a fire has gone out the question "In what direction has the fire gone?" is not appropriate to what has occurred (M I.487). We would consider any answer logically meaningless, but in the context of the Buddhist tradition the question is dismissed because the unenlightened would construe any answers (even those only trying to clarify the situation) as referring to distinct real entities. The problem is not the possible meaninglessness of the questions—there may even be correct answers—but the distortion of the concepts by conceptual projection *(papañca),* a problem central to this tradition's religious quest. So too with any of the alternatives *(koṭi)* connected to the fate of the enlightened: any affirmative answer would introduce a concept—"self" or "a being"—and with it the danger of projecting a real entity. The danger is present in the use of any concepts but would be especially great in any answers to these questions.

The Nature of Language: Ancient Indian and Modern Views

In general, Indian mysticism comes out of a particular tradition concerning language. The various Vedic texts contain the idea that *the word* is itself a power: realities come into being through speech *(vāc)* itself, or speech is the instrument of the creator (*Ṛg-veda* X.7, X.71, X.125, I.64). Interconnections of words came to indicate the interconnections of the realities referred to or the identity of the powers behind them. Sounds and the proper chanting of Vedic verses *(mantras)* came to be seen as important because of the creative and potentially destructive power. One reason for keeping the general public from chanting the Vedas was the real danger seen in elementary mistakes. So too, since they contain the words of creation, the Vedas are older than creation (i.e., are eternal). The *ṛṣis* merely "saw" them (BSB I.3.33). They, unlike the authored traditional works *(smṛti),* are infallible. Many of these ideas are carried on into later Indian thought. The Pūrva Mīmāṃsā viewed language as not man-made but beginningless: Sanskrit words and meanings are not arbitrarily connected, but are eternally and inherently so. So too Śaṁkara claims that all the world comes from the word of the Vedas (BSB I.3.28). The "act of truth" *(satya-kriyā),* which had power only when verbalized, also points to the influence of these views continuing into the classical period.

There is another tendency in both the Theravāda and Advaita traditions, though, that will appear to us as more modern. The Theravādins speak of language as a convention *(sammuti)* (S I.135). Names divide reality into distinct parts, and these parts easily become misconstrued as distinct, permanent, real objects (referents). Language permits us to direct attention away from the here and now to the future, the past, and other places by means of the abstract entities it employs. Hence it is in tension with mindfulness. Yet it should be remembered that lists of categories analyzing all the parts of the experienced world in laborious detail were present from the earliest stratum of the Pāli canon. The Abhidhamma lists were accused of having got out of hand, thereby losing the original intent of the Buddha. What exactly a *dhamma* was and how many *dhammā* there were had become important. But labeling and internalizing the proper analysis of the world was essential to mindfulness: we need to know the true construction of the world in order to see things "as they really are."

For Advaita, the fiction of nescience, *māyā,* arises solely through speech (BSB II.1.27). The world is differentiated by name and form *(nāma-rūpa)* (BU I.4.7). Language becomes central, if not the absolute essence, in the erroneous substitution of a world of distinct entities for reality; we ourselves produce the world of multiplicity (although not reality), that is, the content of our unenlightened awareness is produced by the differentiations we project through language.

Today in the West the view that language is a creative force giving form to reality reappears in a modified form; namely, that different languages codify reality differently, thereby highlighting different features in the flux of phenomena and linguistically directing perceptions. In other words, we create different cultural worlds by parceling up an independent reality into manageable bits through different conceptual schemes. The background-belief for this is that we take language to be a human invention, a social phenomenon consisting of a set of signs, which are arbitrary in the sense that there is no necessary connection between the sound or word "chair" and a chair, and of an implicit set of rules governing the use of these signs. Language by its very nature has a highly practical character, with the assertive function manifest through naming. Each language directs attention along different habitually engrained lines, slicing up realms of experience by the enshrined categories of a society. "The [natural] logic and the [natural] taxonomy contained in a world view are stabilized in the syntax and the semantic structure of the langauge."[4] Different languages are not translations of the same labels; instead, what facts are seen as constituting reality differ. However, languages are not hard and fast screens. Translations are possible (e.g., Greek philosophy into Arabic and then into European languages). Only translating of a doctrinal system's core theoretical terms presents a problem. Translations may be very difficult and the original beauty lost, but the basic ideas can at least be restated. So too, considerably different metaphysical theories of the nature of reality are each statable in the same language. If each language directs attention along certain lines, still major changes are possible: if Euclidean geometry is just an "expression" of Indo-European grammar, still Riemannian and Lobachevskian geometries did develop and prove useful. So also with Newtonian and relativisitc physics: there are major problems with language, but the physicists were not locked into the older point of view.

When scientists or mystics initiate changes in their understanding, they introduce new *concepts* by means of new words or modifications in the meaning of old ones. The same language can be used to express different concepts (e.g., Ptolemaic and Copernican views of the sun), although no set of concepts is neutral to differences regarding what facts constitute reality. Moreover, verbal languages are not the only systems of communication we possess, nor are all concepts expressed easily in a language. Neither science nor mysticism is merely a collection of statements; still, the statements are a very useful avenue of approach for each because of the essential role of language in any thought above a rudimentary level. The concepts in each are verbal in the sense that gestures or pictures are not necessary to explicate them (with the possible exception of models). But they cannot be reduced to their expressions either: theories can be expressed in different words and different languages.

That language is a central force in how we understand the world, perhaps even affecting perceptions, is a point that has been argued by anthropologists, sociologists, psychologists, and philosophers.[5] From Ayer: "There is . . . no sharp distinction between investigating the structure of our language and investigating the structure of the world, since the very notion of there being a world of such and such a character only makes sense within the framework of some system of concepts which language embodies."[6] He continues by saying that this does not mean that there is no world existing apart from our concepts nor that any system of concepts is as good as another. "Even so, our experience is articulated in language, and the world which we envisage as existing at times when we do not is still a world which is structured by our method of describing it."[7]

The role of language in how we see the world does present problems for both scientists and mystics. Concepts have a conservative effect upon our imagination and our perception. Languages do not control absolutely either perception or imagination, but we must internalize alternative conceptual schemes or have experiences that differ radically from the type our habitual conceptual schemes were designed to handle if anything not accentuated by our schemes can be registered in our awareness. All discoveries occur when we can suspend old ideas but still deal with the old problems defined by these ideas. Seeing a new point of view requires a change in the concepts embedded in the language one is using. Each theory dictates what is a natural process and what is impossible in nature. Changing these conceptions is often very difficult. Eventually, at least the ideas in science become basically familiar to the general public.

Vagueness

Another major problem is that, from the modern point of view, all natural languages are ultimately vague. Not only the words central to this project—"experience," "reality," and so on—but all words have degrees of vagueness. The problem here is not the one mystics have in using words that have already-established nonmystical uses, nor is it that many important terms are often used ambiguously. Nor is it that all scientific knowledge is proximate. Rather the problem is that the primary explanatory concepts of any system (e.g., "gravity") are not absolutely precise and clear. In addition, some other concepts in the vocabulary of any enterprise are defined ultimately in circles. Scientists can still manipulate concepts even if their exact sense is never settled—thus the work of scientists in the artificial world of concepts continues. But this vagueness also means that the Cartesian ideal of "clear and distinct" ideas does not obtain even in science.

This vagueness of language in science has an advantage: it aids in imagination and insights. What is perfectly clear in science is likely to be perfectly banal—obscurity points to the depth of science.[8] Some philosophers of science would like to replace these problems of natural language with the logical perfection of an exact, formal language for science. This may not be attainable. As matters now stand, Heisenberg is correct in saying that even "for the physicist the description in plain language will be a criterion of the degree of understanding that has been reached."[9] In the atomic realm, he continues, the problems are as much those of language as those of physics. Pauli would go as far as to say that all the difficulties in quantum theory come from the role of ordinary language.[10] Bohr sees a paradox embedded in quantum theory: it is almost the essence of the experiments used to establish laws that differ from those of classical physics that the observations involved must be described with the concepts of classical physics.[11] That is, in setting up experiments and in describing their results, we go from our own direct experience to our own direct experience, although we are talking about the subatomic realm. Bohr concludes "We have learned that this ordinary language is an inadequate means of communication and orientation but it is nevertheless the presupposition of all science."

The Mirror-Theory of Language

The imprecision of language, one would think, would keep us from basing arguments solely upon concepts rather than upon the experiences giving rise to their use. But the concepts themselves are powerful. We take concepts as indicative of reality, and this easily comes to the point where we no longer take our concepts as convenient groupings of phenomena or guesses at the workings of reality but instead reduce reality to what our concepts denote. More particularly, we project the language expressing our concepts onto reality: reality is reduced to whatever language or set of concepts a culture has as its world-view. That is, to exist is to be the subject of one of our names.

This has two aspects: the level of concepts and the level of structural connections. Quine speaks of a copy-theory of language: elements of language are names of elements of reality, or language is a one-to-one map of reality.[12] And Arthur Danto speaks of Grammatical Realism: that the structure of reality and language is the same.[13] The first type leads to reifying all our concepts: all terms are taken to denote something real in the world. For instance, in some Vedic creation-stories, language creates the world with *naming* being the primary feature of language. In the second variety, no distinction is made between the structure of a language and the structure of the world. A common *logos* makes the world intelligible. From Indo-European languages, we take the world to be a collection of real things (nouns) and events or processes (verbs) that "have" properties (adjectives and adverbs). The normal reaction is to project distinct words into a Humean world of "loose and separate" entities. The opposite reaction is that, because terms operate only in relation to other terms, the world too is composed of relative realities. In fact, no mirroring is necessary at all: we can say something true of reality without any one-to-one mapping. Thus the word "big" does not need to be big to express bigness; and strikingly different metaphysical claims about the structure of reality can be stated by the same language.

The theory that language mirrors reality in a one-to-one fashion is not discussed much by scientists and philosophers of science. Galileo did think that we consider taste, color, and so forth primary properties only because we have names for them.[14] And Hanson discusses how language is not a copy of the reality discussed.[15] Some problems related to "particle" physics may result from a projection of the particle-language upon wave-phenomena. Still, the broader type of a problem (that language and reality share a common structure) is not a major topic. The physicist David Bohm, in an extra-scientific exercise, has discussed "a new mode of language" that will not lead to a world-view of fragmentation, that is, a mode of language reflecting the unbroken and undivided whole flux of reality by giving a basic role to verbs instead of nouns.[16] Such a discussion presupposes the mirror-theory, not attacks it.

It is within mysticism that the problem becomes most apparent. Language, it is claimed, can at best give only a "dim reflection" of the mystical. Mystics perceive reality as not broken up into distinct real entities, and any language operates by making distinctions—that is, by establishing limits and setting the distinguished items apart. Thus for mystics language cannot map true reality in a one-to-one fashion. Languages may not be the source of a sense of permanent, separate objects (there is evidence that infants have this sense earlier than any linguistic ability). But all languages do enable us to enshrine this sense of distinct objects, a sense perhaps necessary at one stage for survival. Thus for mystics the problem is not only that we project ontologically the divisions of some particular language, but that more basically we project *division* as such onto what is either not constituted by distinct objects (for nature-mysticism) or is not an object at all (for depth-mysticism).

During the depth-mystical experience language cannot operate, since no distinctions are present in awareness. The conviction retained from this experience is that reality is ultimately without distinctions and so no language can mirror it. Even when sensory experiences return in the enlightened state, only the being-ness of phenomena is most important. But even "being-ness" cannot apply if we accept the mirror-theory. The situation is somewhat similar to the situation occurring if the whole world were *red:* we would have no word "red" (or "color") because we would have no experience to contrast with the experience of redness, and without the distinction we would not need a word to refer to it. From the enlightened point of view there is only being; there are no distinctions, and so any words even denoting reality from the unenlightened point of view do not apply. For Śaṁkara, many of whose claims entail the mirror-theory, the words "Brahman" and "self" are superimpositions *(adhyāsa)* upon reality. Still, to describe reality without recourse to the differences due to limiting adjuncts is an "utter impossibility" (BUB II.3.6). Even "reality" is a concept distinguishing what is real from what is unreal and thus is a superimposition. Put succinctly, for Śaṁkara Brahman is not an object among objects, while all words arise within that context; how then can words apply to Brahman?[17] Using words opens the possibility that the unenlightened will misinterpret Brahman's status.

Turning to nature-mysticism, we see that the Theravādins attack anything that might become a source of attachment. Language, by separating parts of the flux of phenomena, generates these sources: by following linguistic distinctions, we come to accept our constructs as the reality they are used to deal with. We end up in a false world of distinct, permanent entities rather than the actual world of impermanent, substanceless elements. Any assertive word could be taken to refer to a real entity, and the Theravādins explicitly assert that this is an error. Instead, all such words fall within the chariot-analysis discussed earlier. No substance or real entity corresponding to any assertive word is actually existent. Indicating the conventional nature of our concepts is the philosophical problem of whether replacing a part changes the chariot—is the new chariot the same as or different from the old? What is physically there is not in question, but just what label we apply. All names under the Buddhist conclusion are "a way of counting, a word, an appellation, a convenient designation, a mere name" (S I.135).

This attack against misconstruing the nature of linguistic use is not a blanket condemnation of language. The enlightened Buddha after all did use language. What needs to be rectified is our perception of the nature of the referent, that is, that there are no distinct real entities. Thereupon we see how language actually works. Thus the Buddha speaks without thinking that there are real entities corresponding to his words, or, in his terms to speak without conceptual projection *(papañca)*.[18] He uses "I" and "self" without believing in a permanent independent core to awareness. By means of insight the Buddha is able to employ cultural conventions without getting caught up in the concepts as maps of reality. For him, only the tendency to let the concepts control our sense of reality, to invest our attention in our own creations, had ceased; the ability to use concepts had not ceased. Identifying and labeling in the state of mindfulness should also be viewed in the same light. Similarly, we must not become attached to the doctrine itself, as the parable of the doctrine as a raft only for crossing the sea of suffering (M I.173) indicates. Again, this is not to say that the cognitive content is wrong, but only that right views *(sammādiṭṭhī)* are not themselves real entities that can be grasped and that there is no world of real independent entities being referred to by them.

The enlightened Buddha modified his speech only in the same way that a change in knowledge modifies anyone's use of language—the old words are understood in a new way. The language remained basically the same. A few new terms were introduced but nothing compared to the constant addition of new terms from the advance of empirical knowledge. The impermanent, substanceless factors of the experienced world *(dhammā)* are now seen as the ultimate reality involved in experience. The Buddha's statements then fall into two categories: those of direct meaning *(nītattha),* that is, those about the *dhammā* (which actually constitute reality), and those of indirect meaning *(neyattha),* that is, those about conventional entities such as tables and persons.[19] Statements of the former type need no correction; statements of the latter type may be true or false in their proper context but need qualification with regard to ultimate ontological matters. There is no more contradiction in using the conventional language than there is with scientists treating the earth and our solar system as a dimensionless point for developing some theories: they know the earth in the final analysis is not so constructed, but it is useful for some purposes to treat it as if it were. In both enterprises, claims could be made without the conventions, but it is easier to employ them. Thus the Buddha uses "I" even with his advanced disciples, rather than giving a more complicated and more accurate expression of what is actually there (the *dhammā*). From the Buddhist point of view, the "I" in the expression "I am moving" no more stands for a separate reality than the "it" in "it is raining." Talk of *dhammā* ends this problem, but the unenlightened may still make these factors into independent entities having properties. Thus the danger of ontologically projecting our concepts is not ended by means of direct discourse.

From the Buddhist point of view, the enlightened Buddha no longer names, that is, establishes the existence of real entities by the use of language. With his use of language he remains a "silent one" *(muni)* even while speaking.[20] Silent, that is, because nothing real is mirrored in what he says. This does not mean that such discourse is without *referents* outside the linguistic system or that terms gain meaning exclusively from other terms, but only that such referents (constructed things and the *dhammā*) are not permanent "bits" of reality; they are instead just impermanent, substanceless, conditioned aspects of the flux of phenomena. Nothing "real" (permanent) or related to the destruction of something "real" is involved. It is only the notion that reality is constructed of independent, distinct entities corresponding to or copying our names that is being attacked. Truths can be stated about reality even if there is no one-to-one copying between statements and reality. Nothing in the Pāli texts indicates that statements of direct meaning are anything other than completely true.[21] And to claim that a description necessarily falsifies is to subscribe again to the mirror-theory: the ability to use language may change our mode of awareness from the depth-mystical experience (although not from the enlightened nature-mystical state), but it does not entail that what we say is wrong.

The Theravādins' position can be brought out by contrasting it with the later Wittgenstein's philosophical program. Wittgenstein thought language normally functioned properly (e.g., assertive statements in science); only in philosophical problems was language spinning its wheels without doing its job. When this is rectified, he thought, scientific uses of language would remain the same—everything, in fact, will remain the same except that the philosophical problems will be dissolved. The Buddhists, on the other hand, think that how we see language operating is erroneous for *all* language, *especially* assertive uses (the only ones they discuss in detail). Language will remain the same, but we will not project our categories onto reality. Thus what Wittgenstein and the Theravādins see as the basic problem is totally different: for Wittgenstein, the problem is in only one area of discourse (philosophical uses of language), while for the Buddhists it is basic assertions about objects and the world in any conceptual scheme (e.g., science itself).

For the Theravādins the words themselves and the structure of statements in no way resembles what is referred to, but language is useful in its place. Every language contains conceptual distinctions, but not an ontology describing the ultimate make-up of the world. Only the mirror-theory of reference entails such an ontology. From this point of view, language is still necessary: it directs attention to aspects of the flux of experiences for communicating important matters. But words about experiences should be viewed as color-terms are: all "blues" are not taken to be variations of one underlying blue-ness, nor should we divide the spectrum up into distinct, independent segments. Concepts—especially explanatory ones—are our creations and should be viewed as such: they are vague, proximate, but useful, as long as we remember that reality is not divided up into individual bits. Language is useful for persuasive instruction in the quest for enlightenment also. The Buddha used "skillful means" to lead others, but the danger always remained that, whatever language he used, his listeners might misconstrue its nature and project the categories ontologically. To use the East Asian image, the Buddha's words are yellow leaves taken to be gold.

Silence and Ineffability

The mirror-theory of language can also explain some of the traditional responses of mystics to mystical experiences. Simply put, the central problem is this: language operates by making distinctions and, since the mystical is one without distinctions, it is impossible to copy the mystical with any language. The proper response should be absolute silence. This, however, is hard to maintain in light of the importance attached to the experience: something has to be said, and mystics are often quite verbose. Nor does the problem with language entail that mystics are not vitally interested in doctrine: a proper understanding of the mystical is necessary to every classical tradition. Still, anything said of the mystical is denied. Hence the problem of *ineffability,* that is, that the words used in connection with the mystical actually do not apply.

Śaṁkara on the inapplicability of words to Brahman is an excellent instance of the problem. Brahman is not beyond experience—it can be realized in the depth-mystical experience (self-awareness properly experienced). The meaning of the word "Brahman" is derived from its relationship to this referent outside the language-system. But even introducing the idea of a referent represents to the unenlightened the idea of an object among other objects: we reduce Brahman to an object by the grammatical status of the word "Brahman." And this is precisely nescience. Brahman is permanent, pure, and transcending speech and mind; it is the inner self of all, not falling within the category of "object" (BSB III.2.22). In this passage and in many others, Brahman is characterized as the self *(ātman),* consisting only of understanding *(vijñāna-ghana)* or only of being *(sanmatrā),* and so on, while in the same breath it is claimed that words cannot signify Brahman: words and the mind turn back from Brahman (TU II.4.1, II.9.1), it being unspeakable *(avācya)* and inexpressible *(anirukta)* (TUB II.7.1). Brahman has nothing in common with anything in the realm of dualistic awareness—it has no such qualities, actions, or relations. It is totally undifferentiated while all concepts differentiate. Where understanding *(vijñāna)* is not of a dual nature, nothing can be said (*Maitrī Upaniṣad* VI.7) Even the pronoun "it" differentiates. So too the status of the world is inexpressible since it is not unreal (as is a son of a barren woman) nor real (Brahman, not the pluralistic world); any statement about its status would make the world appear to be a real thing.

In Theravāda Buddhism, on the other hand, the idea of ineffability is connected to what is unexperienceable. *Nibbāna* is not termed "ineffable" or "indescribable" in the Pāli *Nikāyas.* Instead, it is the state of the enlightened after death that we cannot say anything about. This is so because all factors of the experienced world *(dhammā)* are removed, and therefore so are the means of describing the enlightened in that state (*Sutta-nipāṭa* 1075, 1076). In other words, all words ultimately refer only to what is constituted by the factors (cf. M II.66).

Explanations of mystical ineffability usually are insufficient because they overlap with more general problems. Nothing is capable of describing something in its absolute uniqueness: all words involve what we see as common features, and therefore completely isolating the individuality of anything cannot be done in this way. Similarly, all descriptions are inadequate in that there is always something more to anything than the words can capture (e.g., an exact shade of yellow, or the degree of emotion in an experience). The "spiritual blindness" explanation (that we need an experience to know the truth of a statement) applies to all experiential knowledge and does not account for the importance attached to ineffability in mysticism. Another explanation is the logical point that the *that*-ness of reality is ineffable with regard to any concepts because concepts necessarily will involve reference to the *how*-ness of anything: any concepts will involve our categories and our understanding of reality.[22] But this is only a logical point about the nature of language to which anyone could ascribe.

None of these explanations involves reference to the source of mystics' concern with ineffability, that is, the uniqueness of the mystical experiences, or why mystics attach so much importance to the idea. Another explanation seems better: language-use introduces a mode of awareness foreign to the depth-mystical experience; in nature-mysticism, concepts are downgraded in favor of the experience of being-ness. If we have the ability to use language at a given time, then at that time we shall not be able to have the depth-mystical experience. In this sense the mystical is "beyond" the realm of language. But still, why cannot statements *after* the experience accurately reflect something of the experience? Is there nothing expressible from that mode of awareness? In fact, something of the experience does appear to be retained—a sense of reality and nonduality—but there is present also a sense of radical otherness in what is experienced. That mystics accept the mirror-theory of language would explain the unique type of ineffability in mysticism: all terms applied to the mystical are open to the possibility of being taken as signifying objective entities, and so the applicability of all language is denied.

Śaṁkara says that properly only the negation "not this, not this" *(neti neti)* applies to Brahman (BUB II.3.6), since Brahman is free of all distinctions, and negation eleminates all possible specification. Nevertheless, Śaṁkara and other depth-mystics do say positive things about the mystical even while denying it is possible to do so; certain attributes, and not others, are found to be appropriate. For positive characterizations such as "consciousness" *(cit),* Śaṁkara does speak of superimposition *(adhyāsa)* and its negation *(apavāda):* the process whereby a feature known to be inapplicable is applied to Brahman, then recognized as false and driven out by the true idea springing up after the false one (BSB III.3.9). According to Śaṁkara, the procedure is to superimpose upon Brahman attributes that are known to be false in order that no one will believe Brahman does not exist, and then to negate these properties (BGB XIII.13–14). Thereby every positive and negative attribute is denied of Brahman. It is hoped that this will lead us to enlightenment. But if we accept the mirror-theory of language, problems still persist. To invert Leibniz, "all negation is determination," and by this means we who are unenlightened may only separate Brahman from other objects—we again make Brahman into an object. And this procedure does not circumvent the basic problem: why are some assertive words better than others? Why are "knowledge," "bliss," and "consciousness," even if only for the purpose of negating each other, more appropriate then their opposites or other attributes?

The only possible solution is to disagree with these mystics on how language operates and to deny the mirror-theory. That is, the mystical does share some properties with objects of dualistic awareness. Being *one* rather than dual, *real* rather than unreal, and *conscious* rather than a known object are all properties common to the mystical and the phenomenal—even if once we are enlightened, what we consider real and so forth changes. The words "real" and "one" do not change their meaning; only what falls into the appropriate categories does. But because the mystical's ontological status is radically other than that of objects, this commonality must be denied by adherents of the mirror-theory: the reduction of the mystical to the status of an object by the unenlightened is the ever-present danger.

Paradox in Science and Mysticism

This explanation also helps to account for the claims of paradox in mystical discourse. "Paradox" may mean any claim that is out of the ordinary or conflicting with other accepted beliefs. Both science and mysticism are certainly filled with paradoxes in this sense. In fact, science only advances by suggesting what appears at first as absurd. But "paradox" may also be restricted to mean only any set of statements that are logically contradictory. In each case it is statements and thus our *understanding* that may involve paradoxes, not reality in itself.

Does science have paradoxes in the restricted, logical sense? Physicists speak of massless particles, of light bending, of mass increasing with speed, and of additions of speed that never exceed a certain figure. But all of these have reasons behind them that are believed to be sound. The question is whether language discussing such phenomena is ever contradictory. The problem, as noted earlier, comes from having to use language developed for use in the ordinary realm to understand what is nonordinary. Some everyday concepts (e.g., the conservation laws) do apply to the submicroscopic realm, but many do not. Some seemingly paradoxical claims may upon further explication dissolve. For example, to speak of "matter waves" rather than concrete "bits" (particles) of matter requires a reconception of our notion of matter, but the paradox does not lie very deep: once the everyday idea that matter must be discontinuous is made explicit and denied, no contradiction in the phrase "matter-waves" remains. If it does make sense to speak of a moving electron as a "particle," strange situations do occur. For example, an electron can be located at time t_1 at position x and at a later time t_2 located at position y without having moved from x and y at a time between t_1 and t_2. This does contradict our notion of the continuity of motion in the everyday world. But in its own context nothing is logically contradictory. The conflict with the everyday way of understanding only makes it more difficult to understand.

The situation is the same with the Heisenberg Uncertainty Principle, that is, that it is impossible simultaneously to measure as exactly as anyone would wish both the position and velocity of atomic "particles." This is so because (according to the theory) if we were to measure one variable by the only means possible (i.e., electromagnetic radiation), the energy involved would affect the other variable. The more precise we get with position, the more velocity is affected and vice versa. Thus, unlike the everyday realm, where (we assume) measuring the speed of a car in no way affects its speed or other properties, we can measure only one property to any desired degree of precision. But again, this is surprising, but not logically contradictory. Nor is the claim that the measurement itself becomes part of the phenomenon: without our measuring its position, an electron may not have a definite position. We "create" the position by what we do. Not only is this a logically consistent idea, but it is also one not difficult to understand when we realize that in a realm where photons of light are a comparably powerful factor, we cannot "illuminate" something without affecting it.

Problems connected with electromagnetic phenomena as manifesting both wavelike and particlelike properties under different experimental set-ups are similar. A wave-model of light is necessary to explain diffraction and interference phenomena. And a particle-model is necessary to explain the photoelectric effect. Thus light exhibits a continuous nature in one situation and a discontinuous nature in another. To say "light is both a particle and a wave" is indeed paradoxical. It is also wrong: light exhibits behavior in different experiments, which in the everyday realm would be explained by recourse either to waves or particles. What is manifested depends upon our experimental set-ups. That is, what appears depends upon a combination of factors including what we do. We do not know what light is "in itself" apart from our experiments. So far there have been no experiments that can be seen as both wavelike and particlelike; experiments reveal one aspect or the other, not both at the same time. Nothing about the proper description of the situation is paradoxical in the strict sense. In addition, the mathematics of quantum physics produces a consistent account of the phenomena, permitting the prediction of probabilities.

Most scientists are satisfied with the Copenhagen Interpretation of quantum physics, that is, using two limited models to explain the results of light experiments rather than theorizing about light as it "really is" apart from such appearances. This may be merely accepting a limitation on understanding in the place of a paradoxical explanatory posit, and that may be no better than allowing paradoxes to remain embedded in science. Indeed, many physicists (e.g., Bohr)[23] are willing to accept the idea that occasional unresolved paradoxes in the strict sense will occur.

In fact, paradoxes in mysticism appear to come off at least as well as those in science, if not better. They may all be resolvable. Mystical paradoxes would then be comparable to the claim "The sun moves and does not move across the sky" (i.e., the first clause is actually false: the sun does not really move across our sky but only appears to do so as the earth turns). They can be paraphrased in a noncontradictory manner without loss of assertive content, although their soteriological function may be damaged by rendering them more mundane. That is, paraphrasing a paradox may be similar to explaining a joke (i.e., pointing out the points of incongruity where our normal expectations are jolted): it is possible, but the joke is lost. Here the cognitive message is conserved, but its role in the quest for enlightenment may be damaged by making the ontological mirroring of language easier.

Many "paradoxes" in mysticism arise simply from contradictions that mystics are unaware of in their systems. So too some paradoxes are intentional for soteriological reasons. But once belief-claims are made explicit, mystical paradoxes seem to be of the form "Brahman is both *x* and not *x*" where one clause is in fact false. Usually the "paradox" arises from mixing different points of view. For example, the Buddhist belief-claim "there are people, but there are no people": the first clause is made from the ordinary point of view (our ordinary belief that there are people), but is ultimately false (in the final analysis, there are no independent, individual beings but only impermanent *dhammā*). So also "we are separated from Brahman and one with it": the first clause is false, indicating our point of view when we are still nescience-governed. "Brahman changes and does not change (or is in motion and rest)" resolves in the same way. These different points of view differ from those in the wave/particle problem in two regards: (1) In the mystical situation, the different points of view are not of equal epistemological value, while for the scientific problem, both wave- and particle-models have equal legitimacy and equal limitations; and their adequacy for the limited situations is never questioned. (2) Nor is there anything in mysticism comparable to the unifying mathematics of the physics; in mysticism, the problem arises from a dualism of experiences.

A final way of eliminating the apparent inconsistency of some mystical claims involves the mirror-theory of language. For example, "Brahman is both being and nonbeing," that is, Brahman is reality (being), but is other than what we normally conceive as real (and so is labeled "nonbeing"); we may misconstrue the status of Brahman if we apply the word "being" to it. Or "Brahman is nonbeing and not nonbeing" because Brahman is other than what we normally accept as real, but is not nonexistent as is the son of a barren woman. The paraphrases would be rejected in principle only by those who maintain that dualistic language cannot mirror the nondual mystical. The reason for asserting anything arises from the impression that some features of the everyday world express the mystical better than others. But just because the unenlightened will think in terms of objects whenever assertive language occurs (and mystical utterances in general are directed toward the unenlightened), thereby not understanding the status of the mystical properly, does not mean that the paraphrase is false. The claims may be accurate to the enlightened, if all mirroring is rejected.

Language need not be required to copy the structure of what is referred to in order to state the truth. We can use the distinct word "Brahman" without thinking that Brahman is a distinct object among objects. Language is not a substitute for reality but is still useful in its place. All of mystical wisdom may not be expressible in language, but this does not entail that none of it is. Language is the only tool, albeit an imperfect one, for such tasks. And language is one tool flexible enough to reflect back upon itself to see its own nature.

Metaphoric Utterances in Science and Mysticism

These conclusions are reaffirmed when we look at a final traditional linguistic approach to the mystical: the qualified denial of the applicability of terms to the mystical, that is, that language only indirectly indicates the mystical. Analogies and "just so" parables appear throughout Indian religious texts. But the problem runs deeper to *all* words indicating the mystical. Śaṁkara distinguishes figurative from direct uses of language (BSB I.2.11), but also denies that the most direct words actually apply: "reality" (*satya*) cannot denote Brahman but only indirectly refers to it (TUB II.1.1), and even "self" (*ātman*) is qualified by "as it were" (*iti*) to indicate that the concept "self" does not apply (BUB I.4.7).

If mystics were interested in experiences alone and not understanding the mystical, all problems concerning the connection of images to the mystical could be ignored. But mystics are assertive about the mystical and doctrine does play a central role in their lives. The very fact that mystics struggle with their linguistic problems should show that they are concerned with the value of their claims. The metaphoric utterances in particular, since they are the most positive statements mystics make, cannot be ignored.

How metaphors actually work is a matter of debate.[24] Picking out clear instances of them, however, is easier. To show the depth to which metaphors penetrate Indian mystical traditions, even the title "the Buddha" is a metaphor meaning the "awakened one": he is the one who has "awakened" from the "sleep" of nescience to see things as they really are. Such a process extends a meaning (Gk., *metaphora*) from an established situation (ordinary sleep and awakening) to one new and more difficult to understand (mystical enlightenment).

There are different types of metaphoric uses of language: simple metaphors, analogies (which make more explicit the basis and limitations of comparison than do open-ended, simple metaphors), and models (which are more systematically worked out than most analogies) will be of interest here. Some people take all human phenomena to be metaphoric or symbolic, since our constructs are used to stand for something beyond the constructs. Mary Hesse thinks *all* assertive language is metaphoric, since we can generalize only by analogy: no two things are literally identical; classifying two things together highlights their similarities and deemphasizes differences.[25]

Hesse's point may be true. And if so, then of course it trivially follows that all scientific and mystical utterances are symbolic. But noting that two balls are not identical seems quite different from using balls for a model for understanding atoms, where it is known that the differences outweigh the similarities and that only a partial insight into the nature of atoms is possible through the model. In this regard, it is useful to employ another sense of "metaphor", which maintains a distinction between the established ("literal") and metaphoric uses of a term or statement. Whether any or all of scientific and mystical discourse is metaphoric in this sense is far from trivial. The literal use of a term is its everyday use, that is, the use of widest circulation in a group utilizing that language; thus the literal use will reflect ordinary experiences and states of awareness. So too what was once a metaphoric meaning may become literal if it becomes widely accepted and used.

The metaphoric use need not be a colorful image but is any attempt to extend the meaning of a term or statement to a referent other than the established one. Thus one statement may be literal to certain users and metaphoric to others. When the metaphoric meaning is intended, the literal interpretation leads to absurdity or banality. For example, creation-myths are often constructed from familiar sources of creation in a culture's everyday life. The creation of the world is "modeled" upon human birth, spring floods, or whatever. To interpret such a creation-story literally would be to reduce it to a flowery description of a natural or social phenomenon rather than see its true referent (the creation of the world). Similarly with mystical language: the literal-minded will remain within the realm of ordinary experiences from which the ideas used metaphorically first arose. Use of sexual imagery for the mystical experience will be reduced to merely veiled discussions of sexual intercourse rather than recognized as an indirect means of discourse about its true referent (a nonordinary experience or reality).[26]

Metaphors are used in communicating anything fundamentally unusual. That is, some connection to something taken as unproblematic needs to be made as a starting point if understanding is to be possible. Hence analogies and metaphors occur in discussing metaphysical problems such as the mind/body problem and in speculative metaphysical systems built upon root metaphors (a paradigmatic phenomenon extended to explain all of reality).

Metaphors are also necessary for exploring any new phenomenon. The initial understanding of an unusual experience or physical phenomenon will involve noting points of comparison and contrast with familiar phenomena. "Symbol" is sometimes defined in terms of connecting disparate areas of interest. Thus for Berger and Luckmann a symbol is any significant theme that spans spheres of reality by being located in one such sphere and referring to another.[27] Metaphors help organize the way we see the new phenomenon; to speak metaphorically, they give a foothold permitting further investigation. But they also could become restrictive if interpreted literally; that is, the new phenomenon may be reduced to a mirror of the familiar—hence the need when using metaphors systematically to investigate a phenomenon to specify as exactly as possible the exact points of commonality and divergence and also those points which are open to further study. This leads to the general problem of metaphors in scientific uses of language. Scientists themselves form a limited community having a language with its own vocabulary; many terms are introduced by scientists and thus have established (literal) uses in those contexts. But most of the scientific vocabulary comes from the greater culture within which science is set. (Scientific terms also enter this greater society, sometimes with their literal meanings intact and sometimes with metaphoric meanings.) In adapting ideas from the general culture, analogies are used to make the ideas intelligible to the general public and to the scientists themselves, and to give credence to new (and often bizarre) ideas. For example, gravity took on the anthropomorphisms of "attracting" and "pulling." Usually the ideas of physics are relatively simple in their original context and can be explained in quite ordinary language; but seeing their applicability to scientific problems often requires a great deal of imagination. Newtonian space was likened to an empty box without sides or top or bottom: keeping the word "box" meaningful and extending its meaning to space in this situation is not unproblematic. Bohr speaks of the need to employ images and parables in quantum physics.[28] In a recent theory, the "glue" (from which is derived the name "gluon") holding quarks together grows stronger as distance increases; hence, it is likened to rubber bands stretching. Scientists have suggested that conditions in subatomic physics are more like music than architecture: harmonies and pitch involved in symphonies rather than blueprints of mechanical structures better describe the situation.[29] Outside of physics the same holds true; for example, an extraordinary degree of analogy between systems of genetic material and linguistics has been utilized to guide research.

Even more basically, Hesse argues that theoretical concepts in science are always introduced by the metaphoric extension of observational concepts.[30] Because language is from the everyday realm and does undergo transformation in connection with scientific problems (e.g., "straight" Riemannian or Lobachevskian lines are not "straight" in our ordinary Euclidean sense), it is the source of problems when used for scientific purposes. Toulmin can rightly ask why the phrase "three-dimensional surface" in the context of "the universe is a three-dimensional surface of a four-dimensional balloon" is not a contradiction in terms.[31] Space, unpicturable as it is in itself, is likened only to the surface of a sphere; an "inside" or "outside" is not part of the analogy. Whether mystics have any stranger analogies for dealing with the mystical or whether the scientists have any better-defined procedures for applying analogies is questionable. The "language-shift" Toulmin speaks of between a participant's language and the onlooker's language can be acceptable in one realm of discourse and dismissed in the other only if we have other motives or reasons for doing so.

Even philosophers of science are not exempt from metaphoric uses of language. Realist philosophers of science speak of a reality "behind" or "beneath" appearances—as opposed to, say, the workings *inside* a cell. For example, Hempel sees theories as construing phenomena as manifestations of entities and processes that lie behind or beneath them.[32] Mystics too speak of a depth to the world, a depth of oneness unlike the differentiated layer in which science operates; nevertheless, no useful, rigid distinction between uses of language seems possible on this point. Some analogies used in philosophical discussions actually mislead. To speak of "causal *chains*" makes events in the world appear to be separable, distinct links; at a minimum, this reflects a metaphysical position. In addition, the chain analogy does not very well account for how scientists actually work.[33]

Concerning science itself, the most visible instance of its metaphors is a theoretical model. With the loss of visualizability in some physics, models have come to play an especially important role. The phenomena under study in physics can be conceived (faciliated by mathematics), if not "imagined." By making analogies to the unproblematic phenomena, avenues for further investigation open up: the phenomena under study are viewed from a particular point of view, certain questions from the familiar realm are suggested, and so on. This may lead to new theories or, since only parts of phenomena are highlighted, this may actually limit research by not allowing fresh insights. For this, models differ from ordinary analogies by their degree of explicitness for points of commonality and divergence with the phenomena used as a model. Models are usually interpreted realistically: atoms are in some ways in reality like billiard balls. But there are limitations to the analogy and the danger of a literal interpretation of a model is to extend too many features of the familiar phenomenon to the unfamiliar, thereby distorting our understanding of it.

The role of models and other metaphoric utterances deflates a common contrast between religious and scientific assertions, namely, that scientific statements are clear, precise, and unambiguous while religious claims are all metaphoric. Each deals with what is other than initially intelligible and each relies at points upon metaphors, which gain their meaning only in the context of a total conceptual scheme. And scientific metaphoric utterances are subject to the same problem usually used to discredit religious cognitive claims: that the claims cannot be stated in a literal manner. Science does have the mathematical stratum, but understanding still involves more. For religion, the uses of symbols and images to explicate the sacred (e.g., God as a person or a watchmaker) entails the problems involving metaphors in general. But they are aggravated here by the more complex constellation of experiences—especially difficult is the variety of nonordinary experiences—and dimensions of significance. Also, symbols are more pervasive in the various types of religious discourse. Religious images differ from scientific models in some important ways: religious images most often are central to religious schemes in providing the primary conceptions of the sacred (with the explanatory theory of theology playing a subsidiary role); while in science models only connect theoretical concepts to the intelligible, with the other parts of the theory playing at least equally important roles. So too scientific metaphors are more qualified by being tied to particular theories in more exactly specified ways, while religious metaphors permit freer imagination. This increases the danger for religion that the symbol rather than what is being symbolized will become the center of attention. Religious metaphors may last longer than specific theologies (e.g., God as "father");[34] scientific theories and specific models are more closely tied. But although there are these differences, nonliteral terms are a necessary part of science and are not exempt from the problems faced by their religious counterparts.

Mysticism differs from other forms of religiosity with regard to metaphoric language in some important ways. Metaphysical theories are commonly more central than vivid symbols and images; here metaphoric uses of all language present problems. Although mystics form a partially separate community as do scientists, they take over the vocabulary of the religions, metaphysics, and the general culture more than do scientists. Science more than mysticism has a separate language.[35] Some new words may be invented by mystics' communities and therefore meant literally when applied to the mystical.[36] But the great majority of words used in connection with the mystical have established nonmystical uses—for example, "Brahman" had a meaning before the rise of the *Upaniṣads* altered its meaning. Thus there is the problem of distinguishing mystical from nonmystical uses.

The greater mystical problems revolve around the nonordinary type of experience and its object. Plato's story of the cave is somewhat analoguous:

how can the person who left the cave explain to those remaining imprisoned that what they see are only shadows? He would have to use language intelligible to them in order to state something it is not designed for. He could use figurative or nonfigurative terms, but the extension of meaning to another referent would be required. Still Plato's problem is basically only how to explain in a world of two dimensions that there is a third—the depth-mystical problem is how to go from language designed for a three-dimensional world to a reality of no dimension. The mystical is not a being and has no properties: it is utterly simple. No reference to explanatory mechanisms is possible since these are differentiated. The mystical is given spatial symbolism: the mystical is "in," "beyond," "beneath," "behind," "wholly other than," or "through" the world of appearances. So too the true self is the "core," "apex," or "ground" of the ordinary self. Still, all spatializations entail extension and divisions.

How can any metaphors work at all with something void of properties? And why are some descriptions deemed better than others? For Śaṁkara, words do not properly describe or signify Brahman (TUB II.4.1), but imply or direct our attention toward it (BSB III.2.21). "Real" directs us toward Brahman by denying its opposite. By denying the appearances within the phenomenal realm, we are directed away from this realm; that is, by negating all attributes applied to Brahman, all duality, and all change in general, we are directed not to another object, but away from objects. Brahman, not being an object, cannot be presented for identification, but such phrases as "not this, not this" tell us where not to look—an "unknowing" of objects is involved. Śaṁkara's disciple Sureśvara feels that by superimposition Brahman is signified indirectly, just as the statement "The beds are crying" refers indirectly to the children upon them. But this type of suggestiveness *(dhvani)* based upon literal meaning *(abhidhā),* as Sureśvara concedes, only inadequately implies the self, since whatever is used to refer to the real becomes confused with it. It is hoped only that our attention will be directed away from the realm of multiplicity—the context in which words gain their meanings initially—and thereby to the realm of the nondual.

Still, this does not answer the question of how any method would succeed if the mystical is truly free of all characteristics. But it appears that only the mirror-theory of language causes mystics to deny their positive descriptions of the mystical. On the other hand, if we use language without a commitment to the mirror-theory, then "reality" would describe a feature of the mystical. It would be *literally* true if we had a practice established by a common set of mystical experiences; but since "reality" gains its meaning from nonmystical contexts, its mystical use will need to be metaphoric. That is, its meaning is the same as in everyday usage, but its meaning is transferred to a new referent and denied of old ones.[37]

Are *all* statements about the mystical ultimately symbolic? This question brings up Tillich's problem of whether all statements about "being-itself" can be symbolic without a foothold in some nonsymbolic statement. But the problem is that mystics employ terms with established nonmystical everyday or metaphysical uses. At a minimum, for the mystical meaning to emerge, we would have to reject the mirror-theory of reference and all ordinary denotations. Otherwise the mystical would appear to be an *object.* "Cause" would suggest the first cause in a line of causes; "ground" or "being" would make the mystical into an object beneath appearances. Since all our assertive utterances normally refer to some objective reality—something to a degree independent of the individual experiencer and experience—mystical utterances will have to be metaphoric to the unenlightened. The unenlightened must utilize such utterances to go beyond both the normal denotations and the normal tendency to establish the nature of the subject by the nature of the language used. Only if terms (1) without established nonmystical uses and (2) free of this objectifying tendency can be meaningfully introduced will a mystical statement be exempt from this problem. For example, can we understand the phrase "nonobjective reality" to mean that the mystical is real but not of an objective nature? If not, the concrete mystical experience will be the only "literal" foothold.

In addition, only by rejecting the mirror-theory can we make sense of why some concepts recur as descriptions of the mystical and why others are denied. The denial of the mirror-theory vis-à-vis the mystical does not dissolve all the problems of mystical metaphors; it only reduces them to the status of other metaphors. In addition to nonfigurative statements (e.g., "Brahman is nondual"), mystics employ figures to explicate their conceptual systems. The problem is that the most that can be done is to present different analogies to explicate the nature of the mystical—nothing more direct is possible—and that all the analogies ultimately rely upon a duality. The problem for the unenlightened is to be able to move from the multiple to the nondual.

Consider Śaṁkara's analogies. One of his most popular is from the *Chāndogya Upaniṣad* (VI.1.4): for all clay objects, "all modifications exist in name only, being supported by mere words; the clay alone is real." This is extended by analogy to all of reality being Brahman. But the basis of comparison makes Brahman an extended thing and not at all like self-consciousness. Another example is of an unmoving magnet (Brahman) causing iron filings (dualistic phenomena) to move (BSB II.2.2). Here in the basis of the analogy, each part (the magnet and all the filings) has an equal claim to reality, hence a dualism or multiplicity. Śaṁkara must assert that the motion itself is an illusion *(māyā).* In another analogy Brahman is the ocean and phenomena are the waves. This differentiates two parts of one entity (the ocean). So too the idea of an individual as a "drop" of a "universal" consciousness makes consciousness extended and divisible.

Likening Brahman to free, unlimited space in contrast to the space trapped within a jar (BSB I.1.7) again makes Brahman seem to be merely a very large object, not dimensionless. The idea of an unchanging, unique reality is emphasized by the analogy of the multiple, changing images of the single sun or moon in agitated water (BSB III.2.18–20). This too has the advantage of giving some plausibility to the notion of an unaffected and unchanging reality despite our experience of constant change. But it also relies upon duality: the water is as real as the sun or moon. Qualifications remain necessary if we can apply it to the mystical situation. Other analogies relying upon a person's awareness come off somewhat better: the common Indian illustration of a rope (here, Brahman) appearing to be a snake (something other than the rope actually is), the moon (Brahman) not really being multiplied by a person with defective vision (nescience) (BSB II.1.27), or a magician producing an effect and yet not being affected by it (BSB II.1.9). The dualism is reduced to differences in experiences: nescience-guided versus enlightened awareness. The dream-analogy is also useful: the dreamer (Brahman) is "more real" than the content of the dream (dualistic phenomena). All phenomena within the realm of time are wholly other than the ultimately real in this one sense. Still there are problems even here: the dreamer can be affected by the dream (e.g., a nightmare causing one's heart to pound), as Śaṁkara concedes (BSB I.3.42). And the dreamer is an individual, not the consciousness constituting Brahman, suggesting that it is an analogy for solipsism. But this analogy does help us to understand the claim that there are dependent realities and a "ground" to those realities. Still, the "models" Śaṁkara utilized do not fit his "theory" very well. Of necessity, all of them must be drawn from the realm of apparent duality (cf. BSB III.2.11), and none can overcome the difficulty that the same reality constitutes each item involved. This greatly limits the basis for any analogy.

Scientific ideas may be a source of metaphoric images for mystics. For example, one Indian likens Brahman to a huge meson.[38] Less esoteric examples could be given, although each requires a qualification. For example, invisible light (Brahman) through a prism (nescience) produces a spectrum of sensed phenomena; for this analogy to hold, we must think of the prism too as actually the light. So too, instead of talking of the is-ness or the power of being, mystics might view the mystical as a type of energy; but this will remain only an analogy to what scientists discuss (a quantifiable aspect of reality with particular properties rather than the metaphysical "power" common to everything). Still, it permits talk of conservation of energy, a useful law in science, to explicate the changelessness of the nature-mystical whole. Of course, this does not make mysticism scientific any more than science becomes literary by adopting the word "quark" from *Finnegan's Wake*. Mystical claims cannot be proved or transformed by means of the utilization of the ideas of science as analogies any more than by the utilization of analogies from other areas. But such use may enable us to understand the mystical ideas better.

Scientific analogies can also be stretched metaphorically past a breaking point or used in an attempt to give mysticism a more modern aura when in fact they are not providing a genuinely new or fruitful insight. For example, *karma* is sometimes treated as a natural force covered by the laws of energy conservation in order to justify belief in rebirth.[39] But this is open to many objections: (1) Such "energy" is not the type discussed by physicists; an argument would be needed to show the applicability of physical laws to it. (2) If *karma* is an energy that must be conserved, still it need not be in units of persons—or at least the conservation laws do not dictate how the energy must appear. (3) If *karma* must be conserved in units of persons, then there would no way to end the cycle of rebirths—the energy could not be destroyed—and so religions geared toward accomplishing this would be worthless. Thus such an appeal to physics would prove counterproductive.

Those for whom the everyday sense of words is their only legitimate sense may not like metaphors in general. In mysticism, a sympathetic imagination—something perhaps not achievable by all—is necessary to extend words, often vague in their everyday uses, to something extraordinary. But analogies are the best that can be utilized to explicate doctrines on this "path" from one point of view to another.

Mathematics

A note should be added concerning mathematics, it being a highly formal language or set of languages. Mathematical equations consist of convenient signs for concepts, with well-defined formal rules governing their manipulation. On the most elementary level of numbers, mathematics may appear to be perfectly clear and culturally invariant. But even the highly refined language of mathematics is not totally free of vagueness. People were able to manipulate the concept of a mathematical "point" long before it was given a well-defined characterization. Nor, if Imre Lakatos is correct, is mathematics a set of eternal, immutable truths and rigorous proofs built one upon another.[40] Rather there are changing concepts of mathematical truth, standards of mathematical proof, and patterns of growth. Concepts arise and die (and sometimes are revived, as with "infinitesimals") as in the more empirical sciences. Similarly, insights occur in mathematics as in the rest of science. Changing conventions and definitions also play an important role, since mathematics is a matter of convention.[41] The development of non-Euclidean geometries in the last century raised major problems about mathematical certainty. Gödel's work also provided a shock: in any formal system of some complexity, there are propositions that cannot be proved and whose negation cannot be proved; and it is impossible to know that its set of postulates is self-consistent. Axioms in general are not all self-evident to everyone in every culture; some in fact are counterintuitive to us.

Mathematics is open to revision because it is a language designed for particular purposes. This language is not directly tied to experiences but is related more to the logical relationships between our artificial constructs. This entrenches it in the ordinary (dualistic) point of view. Because of its purpose, these artificial languages easily slip into the mirror-theory of reference: the idea of *quantification* makes sense only in a world of sets of isolatable objects, and therefore a reflection of the world in this language may be required for its use. Although mathematical symbols are popular with mystics, mathematics is less applicable to the mystical than ordinary language because it deals with the relations between items rather than with items in isolation, let alone the beingness of such items. Mathematics is basically the opposite of even nature-mysticism, since it deals with concepts without content; distinctions and our mental operations are central.[42] The Theravādins attack any reasoning *(takka)* not tied directly to experience: such reasoning involves the products of our mind reflecting at best upon nescience-guided experiences; it can only conflict with other reasoning and can never substitute for experience.

That mathematics deals with relations among particulars is not coincidental. It arises from the intricate connection of mathematics to empirical science. Measurement and the establishment of mathematical relations, whether universal or probabilistic, have permitted greater precision in prediction, thereby aiding in the development of laws. On the more theoretical levels of science, much of the reasoning involves advanced mathematics. Different systems have been developed (e.g., calculus) to aid with specific scientific problems. "Pure" mathematics may be eternal and unaffected by any experiences (thereby giving rise to the philosophical issue of whether mathematical propositions are *knowledge*-claims), but if so it does not apply to the world of experience. Changes in mathematics do occur with changes in science: the usefulness of some systems is lost and new systems are devised. Any "eternal" propositions are not falsified, but are rendered unneeded. Yet science itself is a human construct that applies to the world only inexactly. And mathematics is a construct one step removed from experience that is utilized by scientists. This fact permits Quine to place mathematics, even in its "purest" form, in empirical science as a whole. For Quine, numbers are theoretical entities—they are posits that are not in space-time (unlike the more empirical posits), but they have the same sort of existence as any posit may have. They are part of the total system that attempts to be smooth and simple, and which squares with perception only along the edges.[43] Such a realist philosophy of mathematics is, of course, controversial.[44]

Mathematics also facilitates science in the way that concepts in general make thinking easier. Nevertheless, applied mathematics is not a substitute for scientific thinking. The fundamental research of science involves new ways of looking at the world. Central scientific ideas need little or no mathematics for their expression, even in the most advanced of theories. Thinking about such unusual realms as quantum levels of reality, however, is often characterized by mathematics: this language permits thinking without visualization or direct empirical content. But the concepts are not simply mathematical equations. Mathematics is the *form,* not the *content,* of our understanding of nature.[45] Refined mathematical correlations are neutral to the issues concerning the nature of the components involved. Thus the coefficients in the equations may remain the same, but the change in the concepts of "mass" and "velocity" from Newtonian to relativistic physics is nonetheless real and in no way deducible from the equations. And through such concepts, as Heisenberg says, the degree of understanding will depend upon "ordinary language."

PART III

Specific Scientific and Mystical Claims Compared

8
Possible Relationships between Scientific and Mystical Claims

From Part II we can see that science and mysticism converge and diverge in different aspects of their nature in a complex manner—matters of degree, not simple distinctions, are usually the case. Each enterprise is a human product involving decisions and arguments rather than being a simple matter of logic and experience. For this final section, the subject will shift to whether selected *claims* themselves that are advanced in specific scientific and mystical systems are the same or not.

Possibility of Common Claims

First some preliminary points. One is that only comparisons between contemporary physics and cosmology and various Asian religious traditions will be discussed. Matters related to psychology and parapsychology have recently also gained much attention. Parapsychology or the technology related to inward observations is sometimes seen as a "clear example of the convergence of Eastern and Western sciences."[1] But a scientific study of mysticism (e.g., recording the physiological correlates of meditators) or a technological aid in meditation is not a reconciliation of *claims* or an illumination of the *nature* of the claims at all. It is not even clear how physiological studies can verify or falsify mystical or other cognitive claims in a simple manner (see the appendix). Second, parapsychological powers are not mystical in the sense discussed here, nor are they in general approved of by mystics. Finally, mystical systems cannot be reduced to psychologies, as is often asserted; they involve a more encompassing way of life and are concerned with ultimate reality, not with the individual self or the mental life alone.[2]

To turn to another point, everything is unique in one sense and everything has something in common with other aspects of reality from other points of view. The problem is to find grounds for insightful comparison. The commonality necessary for comparing the content of mystical and scientific claims is provided by the fact that each enterprise involves conceptual schemes based upon experiences for understanding the world.

This in itself is only very broad, but within this context mystical and scientific claims may conflict or agree. The fundamental differences in basic aims discussed in chapter 3 are not vitiated by the overlap in one abstract area. In a way, R.G.H. Siu's *The Tao of Science* on management is more in keeping with the spirit of Taoism then is Fritjof Capra's discussion of world-views in *The Tao of Physics*. Science will never be a way of life, nor is mysticism merely a metaphysical system.

It could be argued that the differences in approach between science and mysticism are such that claims abstracted from scientific and mystical schemes, no matter how much alike they are worded, must be different in content. That is, the words are the same but the concepts behind them (and the accompanying reasons) remain different. The claims from each enterprise derive their meaning only from their total contexts, and this will involve reference to the fundamental aims of each. Certainly religious concepts cannot be restated in scientific terms simply by substituting scientific terms as translations of Indian or Chinese terms. Showing that the terms are identical in meaning must precede this. The chief problem is how the concepts could be the same if the know-how is so different. Mystics would then have no vested interest in any particular scientific claims about the world. The same situation holds for metaphysical arguments such as Schrödinger's that all consciousness is absolutely one and that all the world is identical to this consciousness.[3] The argument is not in any way based upon mystical experiences but is merely intellectual, and therefore it will not relate to Advaitic concepts or arguments. But the radical incommensurability thesis does not hold for the comparisons of scientific theories because of a commonality of referent (as discussed earlier). And the same is true here, with the referents being such things as reality and consciousness. Cosmological interests in science and mysticism differ: seeing the universe as arising through physical forces or *karma,* or as held together by the four fundamental physical forces or by love or craving; or seeing the universe as dead matter or as a living presence. Science involves structures within the realm of change while mysticism most often involves the arising of the entire realm of change itself. But common grounds still exist for some of the topics: belief-claims can be abstracted from mysticism that may conflict or converge with those of science. The interest in scientific theories may not lie very deep for mystics; but if scientists found, say, any phenomenal permanence, mystics would have to deal with it.

By directing attention only to specific claims, both science and mysticism are truncated in important ways. The total way of life and the quest for enlightenment in mysticism lose central attention. And science is reduced from a way of knowing to merely a static set of current theories. Since theories are often ephemeral, the comparisons can become quickly dated.[4] Theories of what constitutes reality or a human being may change radically in the future as our knowledge and our species evolve. Certainly in the areas of fundamental physics and cosmology, theories of the ultimate physical structure of reality have undergone rapid change in the last twenty years. No one knows how science will develop in the future. Some people believe that we are on the verge of the end of theoretical physics: a new unified, aesthetically beautiful, and basically simple structure of cosmic regularity is soon to come forth.[5] The more usual position is that it is reasonable to expect that a century from now physics will be founded upon unexpected new assumptions replacing currently fashionable ones; science will have a new shape.[6] Mysticism may indirectly contribute to another great scientific revolution, although this is not yet occurring. Or, any such major revolutions in thought may render the specific comparisons now being made between scientific and mystical belief-claims temporary and in the final analysis groundless.

Another type of problem is generalizing about a fundamental "mystical" or "Eastern" world-view and set of values. There are shared characteristics permitting us to speak legitimately of various mystical systems, but this does not mean that all of them share one perennial philosophy or set of doctrines. Certainly Advaita Vedānta, Viśiṣṭādvaita Vedānta, Sāṁkhya-Yoga, and Theravāda Buddhism conflict on some very important points. Thus generalizations that imply that not only all Indian mysticism but all Asian mysticism—and even all Western mysticism—share one set of claims or one world-view are illegitimate.[7] And to this point must be added the fact that even in India alone there have been materialists since before the time of the Buddha who have denied every major religious tenet endorsed by others.

With regard to mystical *experiences* as opposed to *doctrines*, a lack of refined analysis is again present in much current comparative discussion: all such experiences are taken to be of one type. No distinction of fundamental types is made, nor is the possibility discussed that all the experiences within any different types are not the same.

Problems arise regarding science too. For example, a claim common to all sciences and many other conceptual schemes is often taken to have been revealed only in this century by quantum theory or relativity. To illustrate: "the world cannot be as it appears"[8] is not something brand new; that the earth moves and that the stars are not tiny points of light circling the earth reveal the same thing. A major philosophical error is the supposition that what is advanced as a scientific theory is a ready-made world-view. Scientific theories are developed to deal only with specific problems. To expand them into root metaphors for a new metaphysics is to move beyond the science itself into a new type of argument. Using scientific theories as suggestions to develop explanations of other phenomena (as Chew suggests of his "bootstrap" theory)[9] is legitimate, but they need to be worked anew in each realm—or into a metaphysics. Nor is a revolution in metaphysics necessitated by a change in quantum or relativistic physics. From Ian Barbour: "To elevate wave-functions to a central metaphysical status is just as dubious as to make matter-in-motion the ultimate category."[10] Heisenberg more specifically warns against the forced application of scientific concepts in domains where they do not belong, for example, assuming that concepts of quantum theory apply everywhere in biology or other science.[11] The scientists' remarks and speculations are usually highly guarded and tentative, unlike their use by many popularizers.

Many of the comparisons made by writers comparing Western science and Eastern thought are themselves also of a very poor quality. Arguments often of these types: "if two systems have some features in common, they are identical," and "if *A* and *B* see the world differently than do most people, then they see it the same way." Fundamental differences are ignored, as are many important methodological issues. To use an especially egregious illustration: Lawrence LeShan makes many elementary errors in his comparisons of the claims of physicists and mystics.[12] Most important, the juxtaposed claims are not always parallel and their subjects are always different. LeShan rightly says that it is not possible to tell from the content or structure of an isolated statement whether a mystic or a physicist is speaking. This is trivially true: of course the context that gives individual statements meaning is needed—that doctrines abstracted from different contexts may sound the same without having the same concepts is not surprising. It is comparable to juxtaposing the quotation from Hebrews 11:3, "what is seen was made out of things which do not appear," next to a physicist's remark about quarks: no one would suspect that each is about the same subject once the contexts are made explicit. Similarly, when LeShan removes one word from a quotation and asks us whether the "(mystical) absolute" or the "space-time continuum" should be filled in, it does not change the fact that the concepts are radically different: each does refer to what is considered the basic reality and so is similar to that extent, but the depth-mystical is not an object, while the space-time continuum remains an extended, objective reality. The Newtonian concept of an independent, empty space would fit such quotations just as well. Mysticism is still concerned with the oneness of being, science, with the oneness of structure.

Since at least the beginning of this century, there have been attempts to interpret Eastern thought as scientific in nature (although in earlier accounts Eastern thought was considered Newtonian). Why is this considered important? More is involved than an examination of these traditions for possible insights about the world that could be incorporated into a modern world-view. It could be that Western thought is unconsciously or consciously being taken as the supreme standard, with a corresponding lack of sensitivity to other interests: Asian thought must be shown to be positivistic in a time when positivism was in vogue, or existential for those who value existentialism, or scientific today if there is to be anything of worth in it. Or it must share our moral values, if not our beliefs. The various traditions cannot stand on their own terms but must be related to a Western standard. The danger here is in distorting the fundamental nature of these traditions in order to fulfill this demand rather than in understanding them in their own milieu. Such efforts may not be so blatant as that by Indians who claim that the flying chariots in the *Rāmāyana* prove ancient Indians knew aerodynamic theory and had airplanes, but they are of a related sort.

Possible Relationships between Scientific and Mystical Claims

To give specificity to the possible relations between Western scientific and Eastern mystical claims, definite positions can be delineated, each meeting particular requirements.

(1) Compatibility: scientific and mystical claims are compatible if they do not entail claims contradicting each other. If each makes different types of claims about the world, this is not a very stiff requirement or a very valuable point. But reconciling claims that apparently conflict may prove difficult. For instance, *karma* and rebirth could be made consistent with the genetic continuity of parents and children by arguing that the karmic residue involved in rebirth either is totally unrelated to genetic make-up or is automatically directed to the appropriate genetic environment. But if biologists claim that in the final analysis genes are the only factor involved, a conflict will remain.

(2) Parallel or analogy: Components of different conceptual schemes play the same role in the structure of each scheme. For example, atoms as the fundamental reality in one metaphysical scheme versus Brahman in Advaita Vedānta. The isolated items are not necessarily at all similar but merely function in the same capacity. Mere correlations are established. If the items in parallel are also similar in content, something more substantial is established. On the other hand, structuralists might argue that finding a common structure is by itself of significance concerning the nature of reality or of the human mind.

(3) Abstract commonality: some claims have only the common ground of falling into some broader, more abstract conceptual category; once specifics are introduced, they diverge. This property is of limited significance. For mysticism and science, the commonality may be only a feature true of all our conceptual schemes. That science and mysticism are each human products is not very important if the differences are significant. The more limited areas of commonality (e.g., insight-experiences, metaphoric uses of language, or the search for causes) do not change the fact that what is discussed in each is not necessarily the same or that the basic concerns differ. A convergence may be an insignificant coincidence or absolutely crucial, depending upon the configuration of other factors involved. Any mystical "echos" or "anticipations" of scientific theories would be insignificant in this case. Here the danger is claiming that because they each fall into

one broader category they are identical; it is comparable to concluding that a Rolls Royce is a Volkswagen because each is an automobile.

(4) Complementary approaches: science and mysticism are different ways of knowing, each revealing something of reality that the other cannot. Different scientific and mystical claims give a more complete picture only together.

(5) Compartmentalist relation: scientific and mystical claims are separated into autonomous spheres, thereby removing the possibility of conflict (as in the Western Medieval separation of the realms of faith and of reason). Science and mysticism do not approach the same subject, as complements do, but are completely isolated from each other.

(6) Two conflicting types of truths: either scientific or mystical claims are only convenient claims, not the claims that in the final analysis represent our complete understanding of a situation. Ultimately, an error is involved on the conventional level. And so, if each is taken as the final truth, the claims of science and mysticism must conflict. The concepts involved in the claims arise from different approaches, only one of which is supreme, and so they cannot be the same no matter how much the wording may be similar.

(7) Convergence: current science and some traditional Eastern thought are converging upon the same view of the world, either in individual theories or in a total world-view. Science and mysticism are different approaches to reality that lead to the same results. Fritjof Capra, for example, speaks of contemporary science *confirming* Eastern claims.[13] Such a convergence of claims is the most substantial commonality, and the requirements that must be fulfilled for it are the most stringent. Does science or mysticism reveal something of reality that the other does not? Does mysticism provide answers to scientific questions or vice versa? If one enterprise duplicates the other, presumably the more detailed analyses of science would prove more valuable. Related to this is the question of whether science and mysticism are converging in *aims* also, that is, whether the natures of the enterprises are converging.

The realist interpretation of mysticism and the contemporary philosophy of science subscribed to in this project eliminate any position that would reduce mystical claims simply to emotional, ethical, or other value having no entailed claims about reality, or that would give an instrumentalist interpretation of scientific theoretical terms, to resolve any potential conflict. But there may be no final solution to the relation of science and mysticism that could be acceptable to all; different value judgments would necessitate the selection of different positions.

Religion and Science

Looking at the broader discussions of "religion and science" that have arisen out of the contact of Christianity and science helps little in eliminating any positions. This is so because there has been no consensus on the matter. The simplistic position that science refutes religion has in general died, but no one position has arisen to replace it.[14] Every position, from science's being a form of religion[15] to religion's being a form of science[16] has its defenders today. Even a *deus ex machina* of sorts has been revived: some religious scientists argue that Heisenberg's Uncertainty Principle creates a gap that cannot in principle be filled by "natural" forces and that permits God to act in the world through the atomic level.[17]

The position that science disproves religion has usually been replaced with overly facile attempts at reconciliation. Either there is a neat compartmentalism in which science is assigned the job of dealing with nature and religion with morality or a realm beyond the world, or it is assumed that religious beliefs will be modified in light of science. The first alternative ignores the fact that religion must have beliefs about the world as part of the belief-component of the way of life. With regard to the other alternative: if religious beliefs conflict with science, as Christian fundamentalists believe, we cannot demand that the religious beliefs must change—this position would still remain a legitimately religious one (and may not be irrational if these believers can make sense of all "facts" as they see them). So too religious beliefs may be one of the cultural factors influencing what is acceptable in science.

Religion gives a new level of significance to scientific claims (if these claims are accepted) within its total system: it is a religious interpretation of scientific findings. No denial of nonreligious levels of significance is required, nor is there a necessary conflict between the scientific claims and the fundamental belief-tenets of every religion. Such conflicts and agreements are possible, however, when a religious position ascribes a particular purpose to reality that entails certain empirical phenomena and not others. The current debates between "creationists" and "evolutionists" illustrates one of the issues. The creationists have to deny many scientific theories, not just those connected to dating fossils, because evolution holds a well-entrenched position interlocking many firmly accepted physical, geological, and astronomical theories. More basically, there is a conflict of metaphysics—the scientific article of faith that all of reality is lawlike (here, that life has developed free from discontinuities) versus the creationist faith that a willful creator could intervene in nature.[18]

Many religious belief-components are speculative metaphysical in nature—this is especially true of mysticism—and thus neutral to any scientific findings. Śaṁkara, for instance, grants sense-experience and reasoning complete freedom within their proper sphere (the dream) (BUB II.1.20, BGB XVIII.66). In some other religious positions, the entailed belief-claims of the faith are empirical in nature and so may produce genuine agreements or conflicts with science on specific points. For instance, for Christianity in general: Jesus could not have died for our sins if he never lived, and whether he lived or not is a matter of historical investigation.

Even Śaṁkara's compartmentalism falters on the fact that rebirth and *karma* are part of his total way of life; they are empirical in nature (and so in the sphere of science's interest) but still essential to his teachings.

The history of modern science and religious thought involves a series of conflicts. Such conflicts, according to one position, aid religion: this enables religion to purge itself of claims about the world that are not really essential to the faith; a purer faith results. In any case, religion and science, while asking distinct types of questions, do look at the same world. Since the conflicts do appear genuine, the isolationism of compartmentalism must be deemed wrong.

Science and Mysticism as Complementary Ways of Knowing

With the elimination of compartmentalism, a complementarity based upon Bohr's treatment of quantum physics is another popular alternative.[19] For a strict parallel to the complementary use of wave- and particle-models in the Copenhagen Interpretation for understanding a clearly defined problem in physics, quite a few conditions must be satisfied: one common subject; two mutually exclusive approaches, both of which are necessary for the fullest understanding of the subject; resolving seemingly conflicting claims that are both true but limited; and a common theory covering all the phenomena of interest (comparable to the unifying mathematics of quantum physics).

Some physicists have extended this use of complementarity to extrascientific philosophical matters; for example, Heisenberg uses it in dealing with the mind/body problem and with relating the enjoyment of music to the analysis of its structure, and Bohr utilizes it in relating biology to physics.[20] Can we make sense of the relation of science and mysticism in these terms? The common subject matter would be reality; and the mutually exclusive approaches would be the ordinary dualistic point of view (of which scientific experience is a part) and the mystical experiences. In the words of J. Robert Oppenheimer:

> These two ways of thinking, the way of time and history, and the way of eternity and timelessness, are both part of man's efforts to comprehend the world in which he lives. Neither is comprehended in the other nor reducible to it. They are, as we have learned to say in physics, complementary views, each supplementing the other, neither telling the whole story.[21]

The results of the different approaches are the world of "being" and the world of "becoming;" both are the same world, but approached differently. Only when the two points of view are taken together is a fuller picture of reality given. So too the approaches are mutually exclusive: one can attend to the unity of being or to the surface-diversity but not to both at once.

But problems arise immediately. First, consider the dichotomy of two modes of awareness. The ordinary dualistic mode encompasses more than just science. What is the status of, say, common sense in general? And for this mode, is there a mystical experience that is free of any structure and that is a complement? This is possibly true of the depth-mystical experience, but what of the variety of nature-mystical experiences that still have some conceptual or sensory content? Where would they fit into this two-divisional scheme? On the other side, what about the presence of intuitions in science? They may not play a part in the testing of theories, but they are a legitimate part of doing science none the less. Where would this place science as a total enterprise in the complementarity?[22]

A second objection is that complementary points of view are supposed to be equally valid. But mystics, as will be discussed below, would not concede that: the mystical experience has more cognitive significance than does science. And many Western philosophers would take science as more insightful than mysticism. Valuing both equally involves a further judgment and its defense. A third objection is that there is no counterpart in this view to the consistent set of mathematical equations involved with the two models in physics.

In general, the nature of science and mysticism as total enterprises (as discussed in Part II) is not sufficiently different to warrant inclusion in a complementarity-analysis. Only in the matters of basic aims and of which types of experiences are deemed central do they diverge in more then degrees. Such divergence of aims itself leads to problems since in quantum physics one type of question or level of significance is all that is involved. Religious concerns encompass the scientific concerns in a way that is not reciprocated.

Thus quantum physics does not provide a helpful model for understanding the relationship of mysticism to science. It may make sense to speak of mysticism as a complement or supplement in some more general sense of different approaches needed for a full understanding of the world. But why limit the possibility to just two, one general "scientific" and one "mystical" approach? The situation is not nearly so well-defined as that of quantum physics where the two complements exhaust our present understanding.[23] And if science and mysticism are complements only in some vague sense, the same problem of possible conflicts of doctrines arises. In this case, as Northrop says, here is the danger of offering pseudo-solutions to physical and philosophical problems by playing fast and loose with the law of contradiction in the name of complementarity.[24]

Often the Taoist yin-yang sign is adopted in such situations. It is, for example, the center of Bohr's coat of arms. According to this concept, every phenomenon has a "positive" (masculine) and "negative" (feminine) element. But even in their most pure manifestations, there is a touch of each in the other (represented by the dot in each of the two tear-shaped curves constituting a circle). Here there would be two complementary approaches, mystical and scientific, each with a touch or the other in it (unlike the wave-particle model). Only together is our account of reality complete: each account deals with all of reality but from a limited point of view. This is somewhat more flexible than the scientific analogy, but it too involves a commitment to the view that science and mysticism are of *equal* epistemological values.

The Mystical View of Science

If all mystical systems have a common attitude toward science, it is not that science is merely another stance toward reality but instead that science operates in the realm of time; it is concerned only with the content of the "dream" and accepts that content as real, and thus is fundamentally misguided. As the *Śvetāśvatara Upaniṣad* (I.12) says: beyond the knowledge of the eternal, nothing is worth knowing (cf. CU VIII.1.2–3)—knowledge not bearing on liberation is barren. This leads either to a compartmentalism belittling the realm of science or to a position of two conflicting types of truth. As previously mentioned, Śaṁkara grants sense-experience and reasoning complete freedom within the dream since no claim about the content of the dream can be relevant to the reality of the dreamer (the depth-mystical), if not to the Advaitic way of life. The Buddha's general attitude toward science can be illustrated by the example that corpses were used as meditative objects to show impermanence and suffering, not as scientific objects to be dissected and analyzed in order to increase the physical health of the living.

Mysticism contrasts more strongly with science than most religious systems because what is taken to be the fundamental reality contrasts with the realm that science takes seriously: the timeless versus the temporal. As the Medieval Christian mystic Meister Eckhart has said, "as long as one clings to time, space, number, and quantity, he is on the wrong track and God is strange and far away," or more simply, "to be full of things is to be empty of God, while to be empty of things is to be full of God."[25]

The central problem is not that scientific knowledge is irrelevant to the required knowledge and hence a waste of time (mystics have claimed enlightenment long before contemporary science). Nor is it that science is done outside mystical awareness and the scientific attitude is aggressive rather than passive. Nor is it to deny that scientists have found accurate regularities within the realm of time.[26] Rather, the most important problem is that science takes seriously the realm of *māyā:* nescience *(avidyā)* precisely is taking diversity as real: its relation to a reality beyond what is revealed by dualistic awareness is ignored.[27] Such an analytic and objectifying attitude would not only keep mystics who are well advanced along the spiritual path from becoming scientists, it would entail the highest negative judgment regarding science's ultimate claim to knowledge.[28] Any knowledge arising from sense-experience, no matter how accurate, will remain flawed in that its proper status is not understood. Science can never transcend the realm of time and so can never help in realizing the timeless; by its very design, science remains forever removed from the ultimately real.

Science does correct our "knowledge" too; for example, if we want to point as accurately as possible to where the sun really is located, we must point slightly to the west of the image created by eight-minute-old rays. But such "correct knowledge" for distinguishing reality from illusion will remain within the realm of multiplicity. From the mystical point of view, Einstein is correct at least regarding scientific knowledge when he says that we shall never know ultimate reality—"we shall know a little more than we do now, but not the real nature of things." The fact that scientific and mathematical knowledge involves multiple posited constructs used to explain the world of experiences thoroughly embeds it within the realm of discrimination, even if the posits are not taken to be referring to permanently unchanging, independent entities. Discriminations are still increased in number and given centrality in the scheme of knowledge. It is still a world of multiplicity resulting from a misreading of reality, a world produced by our mind functioning without the proper meditative preparation. Reasoning (the manipulation of our constructs) and ordinary sense-experience as the source of data—not mindfulness—remain most important.

Even if the mystics' judgment is accepted, one type of claim abstracted from mystical systems can still be compared to scientific claims. Mystical claims about the relation of the one to the many have no counterpart in science, but claims about structural interconnections on the nature-mystical level may legitimately converge or conflict with scientific ideas. In addition, mystics live in a world of distinctions while outside the depth-mystical experience, and the empirical "science" that mystics would most usually adopt is just the scientific knowledge of the culture of their period, even if there is no impetus in mysticism as such to advance this type of knowledge. A convergence of such classical Indian or Chinese cultural beliefs and Western science would be interesting of course, but it would not fulfill the position that scientific and mystical approaches are converging.

Thus comparing such nature-mystical connections as *karma* and dependent origination to scientific claims will now be the focus of attention. In fact, Capra and others deal almost exclusively with this type of mystical claim. This, however, leads to their chief error: the comparisons involve the dynamic wholeness of some nature-mysticisms but omit the still, distinctionless, *depth-mystical identity*. Thus what is most important to such mystical traditions as Advaita Vedānta is overlooked entirely.

9
Cosmology, Physics, and Mysticism

(1) Cosmology

The issue of the relation of scientific and Eastern claims will be limited to contemporary physics and cosmology. Cosmology is the natural starting-point, since it deals with the largest-scale structures of the universe—their origin and development. In comparing scientific and religious cosmologies, we must remember that the religious claims integrate the physical into a larger picture of reality, thereby covering more levels of significance than merely the physical. That religious accounts of the creation and order of the world are *myths* (stories recounting the human significance of creative events and permitting their reactivation) should also not be forgotten.[1] Some contemporary scientists speak of an abstract "Creator" and of a sort of cosmic version of the teleological argument.[2] But scientific accounts can never have the particular value-component making them myths. Nevertheless, if we focus upon the physical claims entailed by the religious conceptions, a common ground permitting agreement or conflict arises.

Some of the problems involved with religion and cosmology can be seen by the Western responses to the "big bang" theory of creation (i.e., that the present state of the universe results from an explosion of a dense primal mass billions of years ago). Pope Pius XII thought this tended to confirm the Christian doctrine of creation and the existence of God.[3] Or Robert Jastrow sees science as confirming what theologians believed for centuries:

> Now we see how the astronomical evidence leads to a biblical view of the origin of the world [i.e., a "big bang" theory rather than a "steady state" theory]. The details differ, but the essential elements in the astronomical and biblical accounts of Genesis are the same: the chain of events leading to man commenced suddenly and sharply at a definite moment in time, in a flash of light and energy.[4]

The problem is that on none of the "details" do the two converge. There is nothing in the biblical account of a flash of light or energy—light is not even the first thing created. Nor is there an explosion of any kind in this account. In the current scientific theory, at the earliest stage all the matter of the universe (and all the space) was condensed within the space of an atomic nucleus; this was followed by an explosion into a fiery ball of radiation. Nothing comparable—even in a vague "symbolic" sense—is in the biblical account. The two converge only on the broadest items: there was one creation-event at a definite time (i.e., not a steady-state cosmology) and "creation" involved only the ordering of the primal stuff already there (i.e., no *creatio ex nihilo*). Proving there was no creation-event would be detrimental to the literal biblical claim. But a "big bang" is not necessary, and in its features may even be detrimental, since it presents a very different picture from that in the Bible. Nor will any scientific account be sufficient to prove all the religious elements: the question of any purpose to reality lies outside the province of science (as discussed in chapter 3).

If we turn to Indian cosmologies, we see that none play a prominent role in Advaita Vedānta or Theravāda Buddhism.[5] Astronomy/astrology (Skt., *jyotiṣa*) was condemned as a wrong means of livelihood by the Buddha since it is unrelated to the religious concern (D I.12); also this study was connected to astrological forecasting and all such prediction is considered "low art." So too some of the unanswered questions are related to the temporal and spatial extension of the universe. Advaita texts discuss cosmological issues even less than do Buddhist texts. Śaṁkara does say that the world of rebirths is without beginning (BSB I.3.30, II.1.36) and discusses the view that gradually all "selves" will be released from the cycle of rebirth, thereby ending the world (BSB II.2.41). But that is the extent of this issue's relevance to his religious concern.

The *Ṛg-veda* and *Upaniṣads* present various creation accounts. Some may involve *creatio ex nihilo,* unlike the Western accounts mentioned here, since some begin with "nothingness" (*asat,* e.g., RV X.72 or BU I.2.1, although the meaning of "*asat*" is not certain). Other passages stress that there was something (*sat.* e.g., CU VI.2.1). The process of creating the ordered realm usually involves the destruction of something to provide the material—the primal man (*puruṣa,* RV X.90), the golden egg (*hiraṇya-garbha,* RV X.121), or the one (*eka,* RV X.129). The cause is sometimes austerity (*tapas,* RV X.190) or desire (*kāma,* RV X.129). Thus an attempt is made to supply a general mechanism for the cosmogony. Śaṁkara says that the different accounts will agree on the creator (the self), and that the welfare of man does not depend upon the various accounts of creation in any way (BSB I.4.14). For him, Īśvara creates for no purpose other than the joy of doing so *(līlā)*—it is what he does naturally, as naturally as we breathe (BSB II.1.33). The Advaitic concern is more with "continuous creation" in the sense of Brahman's ontological maintenance of all phenomena as the sole ground of being.

How seriously should we accept cosmological speculation here if Śaṁkara and the Buddha did not consider it essential to the religious task? Each tradition merely appropriated its cosmology from the general cultural systems without developing them in any important way. R. F. Gombrich feels that traditional Indian cosmology has nothing to do with Indian science—even astronomy—nor any intimate relation to Indian religion because of the soteriological indifference to the physical universe outside the individual.[6] But there is a danger here of accepting only those aspects of the system which fit our own idea of religious needs. Or even worse, we may accept only what we in the late twentieth-century West accept as accurate as being essential to Buddhism or Advaita, and then argue in circles that they are scientific. Dismissing anything that conflicts with science as "mythological" and nonessential is obviously begging the issue in favor of science as cognitively supreme.

This position does not take seriously the factual beliefs about the general structure of reality—that is, the cosmology—essential to a religious way of life. A legitimate criterion for what the Theravādins themselves take as the factual "essence" of Buddhism is not all the mythocosmology but only the four "noble truths" and the beliefs entailed by them (e.g., a cycle of re-births). This, however, may enclose too little. For instance, Louis de La Vallée Poussin felt the concept of world-ages *(kappā)* is not essential to Buddhism since Buddhism is a religion or a philosophy.[7] But the idea of an infinite number of rebirths seems essential for accounting for the state of a person by means of *kamma,* and this in turn is difficult to reconcile with a definite moment of origin for the universe. An infinite time-frame seems necessary to Buddhist soteriology, and a cycle of world-ages is one way to provide it; the more specialized idea that kammic results can survive the destruction of one world-age to come to fruition in future ones is also necessary for Buddhist soteriology to stand.

Winston King would go farther: Buddhist cosmology has much that is non-Buddhist but as it is found in contemporary Burma it is so congenial to tradition, so useful, and so influential that to speak of it as "nonessential" seems wrong.[8] Its removal would cause a major reconstruction of the total religious structure. For example, the idea of "planes of existence" fits in well with the theory of rebirths, that is, the need for a proper destiny for each reborn being. Other elements of the cosmology reinforce the imperma-nence of things, even if not all cosmological beliefs absorbed from Indian culture are necessary—the earth as flat, India as the center of the world, human life as once averaging 84,000 years long and now growing shorter (D III.84), the calculation of the length of the world-ages (S II.181–82),[9] the distance of the Brahma-worlds above the earth (Mlp 82), and so on.

Even a version of Theravāda Buddhism streamlined by the removal of nonessential Indian accruements is not, as Jayatilleke and others assert, in accord with the findings of science on the nature of a human being and of the cosmos.[10] Some belief-adjustments would still need to be made—most important, the incredibly long cycle of rebirths. The penetration of the social order by the natural order through *kamma* would also need to be eliminated.

Two aspects of Buddhist cosmology are often offered as essentially scientific: the belief in an extensive universe, and the belief in the absorption and reexpansion of worlds. The problem with the first conception, again ignoring "primitive" elements, is that other Buddhist worlds—the realms of form *(rūpa-dhātū)* and the formless *(arūpa-dhātū)*—are not "material" regions in a greater expanse of physical space, but are different "planes" of existence that cannot be reached by physical travel. In addition, at least these other worlds, if not our own realm of desire *(kāma-dhātu),* are the product of our meditative and other kammic activity. There is nothing in science about such a conception of the universe. In general, this conception is no more scientific than medieval Christian ones.

The conception of a universe that periodically expands and contracts fares no better. Its religious significance is to accentuate the duration of suffering, thereby increasing the desire to escape. It may appear to coincide with one contemporary scientific option. But first of all, some beings survive the collapse of the rest of the world and last for many world-ages, the kammic consequences of our deeds requiring more than one world-age to come to fruition. This would directly contradict the scientific idea that everything is destroyed in the heat of the "big bang," permitting no continuity of universes or even knowledge of past ones. And what is central to the Buddhist picture is the role of *kamma: kamma,* not any other physical agents, is the central factor in the creation and destruction of the physical world. The course of the world depends upon the collective individual streams of *kamma.* As King says, it is the cumulative and total *kamma* present in a given moment that holds the universe together as a universe; when this "mental force" decays, the universe deteriorates and finally dissolves until the kammic residue in the higher planes of reality is able to produce a new universe.[11]

It is also a common Indian conception that in each new world-age, individuals bear the same names and forms as the individuals of the preceding world-ages: worlds are arranged exactly the same in all world-ages (BBS I.3.30). The same beings and events recur over and over again with different "persons" (i.e., different karmic streams) filling the roles. The Buddhists do not share this belief totally since they speak of world-ages void of Buddhas *(suññakappā).* But the idea of a pattern which is repeated over and over—an "eternal return"—is certainly contrary to scientific theories in whose conception even the *laws* of nature may not repeat.

In another way the Indian conceptions of the world-ages differ from the scientific: there is no conception of the continual movement (expansion or contraction) of the world. In the West, this is a new idea. Copernicus's view, for instance, is of a static universe with the sun in the center. It was only in this century that the idea that the content of the universe is expanding was advanced. Einstein and many others were reluctant to accept it at first. The Indian conceptions are of an "unrolling" *(sṛṣṭi)* and "absorption" *(pralaya)* of the universe, but these do not involve any motion once the content is "unrolled."[12] Nor is the decay of the natural order occurring after the golden age *(kṛta-yuga)* ever described as involving the world's collapsing together. Rather, there is an emergence ("creation") from an unmanifest state and dissolution back into this primal state, not a contraction into a primal atom.[13] Part of this is the decline of the social-nature order (human life getting shorter, and so on), something conflicting with the scientific optimism that technology will improve matters (until the final destruction of the universe).

Many of these latter points may not be essential to the religious quests of the Theravādins and Advaitins, but they show how quickly the Indian ideas diverge from scientific ideas once any amount of detail is introduced. More important, these ideas are not scientific in essence, unless "essence" means merely the broadest outline void of any explicit content. In that cause, its essence converges with science only in the same way as Chirstian temporal linearity converges with other scientific alternatives.

(2) Physics

Relativity and Relativism

As discussed in Part II, there is no conception in classical India of space and time combined or of either time or space as an especially fundamental reality. Buddhism and Advaita are about all of reality, and nothing about astronomical scales and structures is of central importance. But relativity requires mention because it is often equated with a lack of objectivity in any sense or with the idea that the world is a "dream." Like complementarity, it is used to justify any relativism. For example, one Indian commenting on the difference in the length of a human's and a god's life says that according "to the Theory of Relativity time is relative and one second may be a thousand years."[14] But the relativity of time involves only the velocity of any object (sentient or insentient). The basic difference between relativity and Newtonian phyics in this regard is the need to specify the framework within which measurements are made: there is no absolute motion, but only motions of objects relative to other objects. That we cannot tell which object is moving solely by reference to motions was first advanced by Nicholas of Oresme in the fourteenth century concerning the relation of the sun and the earth; it was not a revelation of Einstein.[15]

Nor does relativity mean that there is nothing absolute. The space-time interval between two events is fixed regardless of framework. Or, for the "twin paradox," when the astronaut twin returns there will be a real difference in appearances. How much time has passed since their birth will have no absolute answer, but the "clock" of each twin will register only one reading. With a sense of acceleration, one can tell with certainly either that one is moving or that gravity has suddenly increased greatly. But something real has happened; there is more involved than just a difference in points of view (frameworks). Thus in no way can we equate relativity with a relativism. "Einstein had devoted his life to probing that objective world of physical processes which runs its course in space and time, independent of us, according to firm laws."[16] And relatively does not change this.

Fields and Mystical Oneness

Science since the time of ancient Greek speculative thought has had a monistic tendency. That is, independent concepts have been replaced by concepts connecting items, thereby reducing the number of independent concepts. Thus electricity and magnetism are now seen as two types of the same energy. Space and time are not related in that manner but are no longer seen as existing independently of each other. Matter and energy are seen as two forms of one reality. Matter, like light, has particle- and wave-manifestations. Matter-energy in turn is considered merely the curvature of space-time, according to one theory. Space-time becomes a field, with matter being concentration-vortices of it. That is, a particle is a condensation of the same "stuff" as the field; it is a bright red spot in a pink medium, as it were. Space-time becomes the primary reality, the fundamental field generating matter, with the matter itself secondary. There is no plurality of independent real entities—all are connected to the primal ground. The ultimate "stuff" of reality is the continous whole of space-time.

This last view has more in common with nature-mystical wholeness than with a Newtonian empty independent space populated by distinct entities operating upon each other externally. It is comparable with the wave/ocean analogy used in mysticism to overcome the idea of any independent real entities and to emphasize the common ground of reality. But this agreement cannot be exclusively emphasized. Even if nature-mysticism is explicable by the wave/ocean analogy, still nature-mystics do not *see* scientific fields. Rather in mystical experiences, there is a blending of objects in the sense that boundaries are less noticed in the light of impermanence and the common experienced being-ness. Most important, this being-ness is not an abstract category but a felt, concrete power.[17] But the scientific constructs such as "fields" remain unperceived cosmological posits not changing our experience (which occurs on the everyday level), for not even a Gestalt-like switch occurs in our perception while mysticism involves precisely this change in experience. The scientific analysis here cannot even be said to provide an explanation for what is experienced, if the difference in mysticism is with experience while the scientific theories are neutral to any experience. Nor, for this reason, can scientists be said to have reached by theoretical reasoning what mystics have experienced. Nor does science *confirm* the mystical claim insofar as the impermanence and beingness are compatible with the empty Newtonian space—a conception of space comparable to the Indian concept of *ākāśa* (as discussed earlier). There is nothing more "spiritual" about the newer conceptions, nor can they be related any more directly to such soteriological problems as suffering.

And finally there is the depth-mystical—a reality free of any differentiations, extensionless, timeless rather than eternal, and nonobjective. It differs from space-time in each of these regards. Mysticism focuses upon the identity of *being* while science focuses upon *structures*. Space-time has the structure necessary to explain why one specific cosmological state of affairs is the case; the depth-mystical identity is structureless and its relation to any structured unity of parts remains a mystery.

Substance and Voidness

With regard to another aspect of the same situation, science and mysticism both search for some unchanging "substance." Space-time and also quarks can be seen as the current results of this process.[18] But it is also true that impermanence is prominent in current science. The idea of solid atoms evolved into that of subatomic particles interrelated almost organically, with none of the particles being primary. At a minimum, at the level above any possible "bare" quarks, there is no unchanging substance or particles. And if quarks are "bits" of matter, still, according to one theory, they are not independent of space-time. Quantum phenomena are created (not released) and destroyed in interaction-events—blinking in and out of "existence," as it were. No one phenomenon is isolatable from the others. Structures and patterns show more permanence than what "fills" them.

Mystics too speak of a lack of substance, although on a macrocosmic level: a sense of self and other real entities is a distortion of the actual nature of reality and the actual connections of parts. Is Heisenberg then on solid ground when he muses that there is perhaps an affinity between Far Eastern tradition and quantum theory, since the former has not gone through naive materialistic thinking?[19] More exactly, is the denial of substance in favor of energy and motion in contemporary physics identical to the theory of no-self (*anattā*) in the Theravāda tradition or of voidness (*śūnyatā*) in the Mādhyamika tradition?

Both Buddhist traditions, like all Buddhism, argue against permanence, but their foci are different. The Theravādins emphasize the fact that all constructed things (*saṅkhārā*) are impermanent, and they analyze them into the factors of the experienced world (*dhammā*) constituting them. The

Theravāda Abhidhammists found, for example, that matter (*rūpa,* i.e., earth, water, fire, and air) could be analyzed into 28 such factors. These factors in turn are substanceless (*anattā*). But the chariot-analysis is the paradigm of the Theravāda's system; it shows that there is nothing on the everyday level to which we can become attached. The Prajñā-pāramitā and Mādhyamika Buddhists in reaction to the proliferation of *dhammā* in the Adhidhamma systems shifted the focus to the lack of substance in a radical statement of the no-self doctrine: there is absolutely nothing to individualize existence—everything is void of self-existence (*svabhāva*). This is not to deny that there are *dhammā,* but only to focus attention on their impermanence.

Although the Buddhist ontology is one of impermanent basic building-blocks, these *dhammā* do not correspond in any way to atoms: they are directly experienced, not theoretical, and are more like qualities than substances. Their size is not an issue—the smallest particles are analyzed by the Theravādins into points corresponding to the directions of space. They are parallel only in fulfilling similar functions in different conceptual systems. Mystical insubstantiality has nothing to do with empty spaces on either an atomic or cosmological level, or with energy rather than matter. "Voidness" did develop eventually into an ontological absolute reality, a ground of "becoming" characteristic of nature-mysticisms. But in no Indian or Buddhist tradition is there anything comparable to a notion of a tendency to exist, or to anything going out of existence as a particle to become energy, or to anything coming out of or returning to a ground of becoming in this manner. "Form" and "voidness" in the *Heart Sūtra* involve the lack of *permanent* entities, not particles arising from underlying *fields,* as Capra feels.[20] The subject is the conditionality between elements of constructed entities. Even the impermanent constructed things might last for many years (and kammic fruit might take more than a world-age to come to fruition); there is no conception of change in every moment, with its problem of how continuity is possible.

The Buddhist concept is our everyday notion of impermanence applied more systematically and thoroughly to everything. A cosmology with the theory of a destruction of everything has more relevance to the Buddhist concept of impermanence than does quantum theory: only this would rule out the possibility of macroconfigurations' lasting forever. Would the Buddhist no-self doctrine be falsified if the existence of unchanging bits of matter, a space-time continuum, or some other scientific core of permanence were established? Yes, for there is the entailed belief-claim of no permanence with regard to all reality. If some unchanging reality were discovered, the impermanence of the everyday realm would nevertheless remain untouched, and thus the truth that things constructed of the bits of reality or the changing parts of the conserved whole are impermanent would remain. There would still be nothing within the realm of direct experience to which we could become attached. And the problem of suffering would still remain.

This topic must also neglect the permanence of the depth-mystical. Śaṁkara's fundamental "substance" (awareness) is nonmaterial but permanent. And it cannot be reduced to any permanence within the "dream."

Interconnections and Conditionality

Another topic concerns the patterns of relations and connections that are the ultimate subject matter of science. The mechanistic approach to nature—cutting reality into independent bits and analyzing the relation of one bit to the next—is still most widely accepted today even in areas such as molecular biology. But system-analyses are becoming popular in quantum physics and relativity: the undivided whole is the principal reality, all parts are in flux and dependent upon the whole environment, and "particles" are our abstractions from this whole.[21] The self-maintaining, mutual adjustment of parts within the complex whole is likened to that within an organism.[22]

Is the nature-mysticism of the Theravādins at all like this? This tradition is minimally compatible with an organic/holistic approach, since impermanence and conditionality are compatible. But it is also compatible with mechanistic world-views. The chariot-analysis could just as easily be of a *watch,* the paradigmatic root metaphor of mechanistic metaphysics. Aspects of the Theravāda world-view conflict with a thoroughgoing holism (e.g., space being independent of the activity in it). And no part of the Theravāda analysis is strictly incompatible with mechanistic metaphysics.[23]

While being compatible, Buddhist impermanence is not itself an interactive holism. The reality as a "whole" is not discussed at all—not even as an unanswered question. If the Theravāda were to discuss it, they might claim it is impermanent (to remove a possible object of attachment). In the chariot-analysis, there is the claim that the parts arranged only in the proper manner have the effect of being a chariot. But the point of the analysis is that there is nothing real corresponding to the concept "chariot"; there are only parts, and the parts are themselves without substance. The wholeness and the relation of the parts are both clearly secondary. Nor is the emphasis on *becoming* in the Theravāda comparable to the integration of parts in a process within a whole. The emphasis is simply upon nonindependent parts arising and falling in an orderly fashion. How they are related to produce an effective whole is not discussed. To shift attention to that matter would prove to be a hindrance to the religious path since permanence of structure (and perhaps of the whole) would become the focus of attention. Certainly there is nothing on the process or relations alone being real. In the Theravāda analysis, any event has an actor, an action, and an object of which none is real; only the substanceless *dhammā* are real. It is an ontology of parts, not of events or of relations.

The Prajñāpāramitā and Mādhyamika traditions' emphasis on the *dhammā* being void of self-existence agrees with the Theravāda's on this point. Nāgārjuna's method is to shift attention to the concepts (e.g., interdependence of concepts) in order to show the inapplicabililty of any of them to reality: if there is no "actor," then there can be no "action," and so on.[24] His most basic argument in this regard is this: if any process is possible (and experience shows us that there are processes), nothing real (permanent and unchanging) can enter into a process, and if no real entities are involved, then there is no real process either (for if there are no real components, how can the event be real?). If nothing is real, nothing can be truly different or identical. All that is left is the rise and fall of dependently arising *dhammā* void of self-existence. There is once again the emphasis upon conditionality and impermanence—that things arise and fall in sequence—but nothing on the mutual adjustment of parts constituting a stable whole. Nor is anything here comparable with the idea that "all properties depend upon all other entities." Things are in flux but not reacting back and forth as in an organism.

The Buddhist tradition most often considered holistic is that related to the *Avataṁsaka Sūtras.* Its world-view is described by D. T. Suzuki as one of interpenetrating lights. The knowledge of these texts for the writers on Western science and Eastern religion seems to be limited to the analogy of Indra's net, a net of jewels so arranged that each gem reflects all the others. Thus all the jewels are "in" one and one is "in" all. Applied to the world, the whole universe is "in" a particle of dust. There is a nonobstructed interpenetration, not of the absolute and phenomenal (although this is also important in these texts),[25] but of phenomena with phenomena. According to Suzuki, with such interpenetration "each individual contains in itself all other individual existences as such."[26]

But what is meant by "interpenetration" and "contains"? Is it not a contradiction to say that a part can literally contain the whole of surface-phenomena or anything extended? If Chang correctly explains the concepts, "mutual *interpenetration*" means only that everything *conditions* everything else: everything conditions and is conditioned (or mirrors and is an image).[27] And *"contains"* means merely *interdependence:* "A contains B, C, and D" means A arises dependent upon B, C, and D; B arises dependent upon A, C, and D; and so on.[28] The multiple well-defined gems are there, but no self-existence. Again it is the dependent arising and the lack of individualizing substance that is being stressed, but with a new twist: the mutual interdependence. For other traditions, A depends upon B, but not vice versa; there are repeating chains, but not immediate circles. The problem this latter view raises is that apparently each entity is constituted only by relations to the first and other entities. The relations or structures are not discussed as real, and the nonindependent centers seem to dissolve. We have mirrors mirroring other mirrors, and, one might ask, where do the images come from?[29]

This is definitely more holistic than the Theravāda and Mādhyamika points of view. Still, there is no way that the parts and the spatially separate areas can be denied. We are not present in every point, but are merely interconnected parts of an undivided whole. Distance and travel are not illusions. As Susuki says, space is still extension[30]—there is no contraction into a point. The north and south poles of a compass do not lose their characteristics, but must be seen as gaining these properties by interconnections of some sort. Everything is not absolutely *identical* with everything else, literally present in the tiniest and largest parts of space. Rather we are one with the undivided whole by the lack of absolute independence. Each entity is conditioned by each of the other entities within the whole, but not necessarily in a one-to-one manner, that is, direct contact; as parts of one segment of the interacting whole, the segment or the rest of the totality as such may produce effects. We see interconnections by the "nonobstruction of the mind," not by science with its differentiating and objectifying concepts or any other ideas while we are still unenlightened.

Mach's idea that the inertial mass of each entity is determined by the whole of the universe and contains information about the whole universe—an idea predating quantum physics—is the same broad type of idea. Other more specialized theories have something in common too. One is Geoffrey Chew's "bootstrap" theory of strongly interacting particles (hadrons).[31] It is similar to Avataṁsaka views. Here no elementary particle is fundamental, but each produces each other through interactions. The "self-consistency" Chew speaks of is that each particle contributes to the creation of other particles, which in turn create it. Each hadron "contains" each other one, not literally, but in the Avataṁsaka sense. But unlike the Buddhist theory which covers all of reality, bootstrap-theory is very limited. It cannot even account for all the phenomena on the level of reality's organization for which it is designed: weak interacting particles (leptons) and photons of light are not encompassed by the theory. Equally important, only within limited systems on one level of organization does this interaction occur, not on an atomic level or above. Nor is there any suggestion that the universe as a whole affects each part or that the hadrons on earth affect, say, the hadrons in a distant star.

The basic difference in nature between any of the recent scientific theories offered for comparison with mysticism and the mystical claims is that in science the whole is treated as objective and the differentiations (the lawful structures) are the focus of attention. There is also a fundamental difference in scope: mystical wholeness involves all of reality, especially the experiential level, while scientific theories deal only with very limited specified ranges of phenomena on subatomic levels. Expanding the scientific theories into metaphysics by means of analogies (with its accompanying problems) would cost at least the mathematical refinement of science, if not more. That the theories of the subatomic realm are neutral to differences on the everyday level precludes any greater convergence than abstract commonality or any scientific explanations of mystical mindfulness.

Holography and Reality

A recent technological achievement connected to field-properties, holography, is also advanced as important here. A hologram is produced by two beams of monochromatic light, one aimed at a recording plate and one bouncing off the object being "photographed." The interference-pattern of the two beams is recorded on the plate, and when the plate is illuminated by the same type of light, an image of the object in three dimensions is produced if viewed from an angle aligned with the original wave-front. Its feature most relevant to the issues at hand is that we can illuminate only a tiny portion of the plate and still produce the entire image with its structures. That is, each fragment contains the information of the whole. This idea is now being applied to the problem of memory and other mental functions that do not seem localized (e.g., we can surgically detach any part of the brain without a loss of memory): past events are the objects, memories the images, and the brain is the recording machine and plate (with neurons or brain-waves as the coherent light-source).

Bohm proposes a similar order implicitly contained in its totality in each region of space-time.[32] Can this idea be further expanded to explain the world as a whole? Or can it be applied to the Avataṁsaka world-view of "one single light [which] reflects in itself all other lights generally and individually?"[33] The requirements are strict: the detailed aspects of the holographic process must be the same. In the most important way, this process is the direct opposite of mystical reality: a hologram, like a memory, is an empty image, while mysticism involves *being,* the reality underlying the "dream." The former relates only to structures, while the latter is concerned with the ontological basis, that is, getting behind the "illusion" of taking the structuring as indicative of the nature of reality. "To see the world in a grain of sand" is seeing either the beingness of all reality or the interrelatedness (conditionality) of all reality; it is not seeing the entire structure of the universe. In holograms, it is all structural information that is present in each fragment. The mystical view is that the totality of reality is "in" each part—the moon is reflected *in toto* in each drop of water. Reality is "encoded" in each fragment, no matter how small, while the information encoded in a hologram loses clarity by degrees as the fragments become smaller; hence when a level too small to record wave-interference is reached, the holographic information, unlike the mystical reality, is lost. Thus, even if reality were holographic (containing all the hierarchically arranged structural information in each fragment, no matter how small), still mysticism is concerned with the opposite—being, experienced either as what all phenomena have in common or as the utterly structureless depth-mystical.[34]

The idea that the world is a "thought in the mind of God" is certainly not new. Does holography provide a mechanism that would give credence to this idea? First, we would have to ignore the question of what corresponds to the coherent light, the mirrors, and the plate: reality must be viewed as waves existing independently of us and being structured by a fragment of reality (the brain, the "harmony of the spheres," or whatever). Thus the equipment for the hologram is part of the hologram. In addition, there is the major problem that holograms and memories merely reproduce images of real objects. For a parallel to holography to hold, our world would have to be an exact copy of another real world. This is not a mere detail but something central to the holographic process. Unless a parallel substantial world is accepted—with the problem again arising of where it came from or what its relation to the undifferentiated ground of being is—introducing holographic technology only complicates, not clarifies, the situation. Thus this may be a twentieth-century version of the analogy that the world is the wave-phenomena of an ocean of pure awareness, and no more or less explanatory. In the end, it appears that holography cannot be used convincingly as a model to account for all of reality.

Consciousness and Matter

A very different type of holism attempts to integrate awareness and the physical. This is not a materialistic monism, or a mental monism such as Śaṁkara's or Schrödinger's. Rather, the holistic approach here emphasizes that the observer, too, is part of the world: observation is by part of the world, not just *in* it. Observation is participation of the observer and the observed. But for a holism to obtain, the act of perception must *affect* the object of perception and vice versa: there needs to be an interaction possible in both directions. How awareness can interact with the material is hard to understand; the idea that awareness generates brain-waves that interact with waves of matter in the universe does no more than Descartes's theory of the pineal gland for understanding how the process actually begins.

There are some errors committed by those writing on a matter-consciousness holism. First, nature-mysticisms such as Buddhism do not discuss the interaction of perceiver and object in perception: the perceiver, each act of perception, and the object of perception are analyzed without asserting that consciousness affects the object or vice versa. The most usual Indian view is that the mind modifies itself to the shape of any object in an act of perception—no interaction is involved.[35] Obviously, "participation" is an inappropriate description in this regard. The mystical interest is in whether what is seen is real or not—*knowledge,* not control or effects, is what is crucial. Nāgārjuna's analysis illustrates the mystics' approach: if there is no real actor (no independent, permanent agent), there can be no action and hence no object of action either. Only with paranormal powers is the physical affected by the act. And how this is to be explained is not a major subject. More important to mysticism are the disinterested observations of mindfulness.

Second, the most common scientific example of the influence of consciousness upon nature—the Heisenberg Uncertainty Principle—does not involve such influence. The interference that precludes ascertaining simultaneously both the position and the velocity of a subatomic particle results from the sensory equipment, i.e., from energy that is as physical as the object, rather than from anything mental. The measurement becomes part of the process: the *radiation* needed to "illuminate" the event enters into the event itself, and any attempt to sharpen the picture only increases the disruption. But there is nothing about *mind* affecting matter here. We "create" phenomena by our measurement, and do not know what occurs between our measurements or independent of our measurements. (Whether this limitation is because of the nature of reality or is due to the limitation of our capacity to know is a matter of debate). This sort of "participation" is not so unusual: any experiment involves manipulation—something appears that would not have appeared without it. When Heisenberg says that what we observe is not nature in itself but nature exposed to our method of questioning, he is saying something true of all science, although he is thinking only of recent physics.[36] Even perception "creates" sounds, colors, and so on, since these result from an interaction between an observing subject and an object. Quantum physics differs only in the interference of the measuring-equipment—nothing about the subject and object changes.

There is a controversial theory that consciousness is the "hidden variable" that determines the collapse of the wave-function (e.g., the factor determining which possible course of a photon in a two-slit experiment is actualized).[37] This would be an interaction of mind and matter resulting in an effect upon matter. The distinction between perceiver and perceived would no longer be absolutely clear, but the two would not be identical either.[38] This holism would be very limited, however, if this is the only possible effect of consciousness upon matter that scientists are willing to consider.

None of these suggestions for a mental-physical holism compares with the creation by awareness in nature-mysticism. Nescience, our role in making "objects" (i.e., a plurality of distinct entities), applies to all of the phenomenal realm, not just to some areas of it. An epistemological misreading, not a causal interaction, is involved. The Theravāda idea that at least some realms are the product of consciousness also differs: this is creation of all phenomena in a realm, not one factor determining only selected events therein. Causing a world is different from causing only a limited number of events *within* a world (the other events supposedly occurring without the aid of consciousness).

Certainly the depth-mysticism of Advaita Vedānta, for which consciousness is the only reality, and not merely one of two types of powers integrated into a whole, is radically different. In general, the depth-mystical is different from any holism having parts: "whole" and "parts" cannot be applied to Brahman, which is acknowledged to be without parts (BSB IV.3.14). The dreamer is totally "in" each part of the dream: any holism would be within the dream-realm. The relation is one of *identity*, not a connected, extended *whole*. As Ian Barbour says, mystical undifferentiated identity "is very different from the organized integration of cooperatively interacting parts that characterize the unity of a body."[39] None of these holisms help explain the relation of the timeless, spaceless depth-mystical to the phenomenal realm; the "parts" would take on too much reality, no matter how unreal they are in relation to the depth-identity.

Submicroscopic versus Everyday Realms

A primary problem throughout all these topics is the issue of the applicability of theories designed for submicroscopic (or cosmological) events to the macroscopic level of the everyday world. The many different levels of organization to reality have a great difference in their effects on the levels of hadrons, of cells, and so on. How do theories from different levels apply to the ordinary macroscopic level? The question is not how theories explain ordinary phenomena, nor how quantum or cosmological events may effect or even produce this realm, but how theories devised for dealing with problems on one level can be appropriated for another. Do such theories force us to view the everyday realm differently? Do they logically compel us to deduce certain claims about the macroscopic order?

The problem with affirmative answers is that the inverse of the problems with language in the quantum realm arises: if we cannot apply concepts from the everyday realm to the submicroscopic realm without severe qualifications, then the reverse is also true, that is, the descriptions of the submicroscopic are inapplicable in the everyday realm.[40] Concepts from one realm cannot glibly be applied to the other without an argument for their applicability derived from the phenomena of the realm in question. The alternative would have to rest upon rather tortuous reasoning: (1) Concepts designed for the macroworld do not apply to the microworld; (2) only the microworld is real (this being a metaphysical belief); and therefore (3) concepts designed for the microworld must apply to the macroscopic everyday world. To give some obvious examples of the inapplicability of submicroscopic concepts to the everyday world: planets do not jump out of their orbits in sudden, discontinuous shifts as electrons appear to do. Nor can we argue directly from Heisenberg's uncertainty principle to the conclusion that the act of perception affects trees and other such objects, or that we cannot measure simultaneously the velocity and position of a car or a planet to whatever degree of accuracy we choose without affecting either in any way; in the everyday realm we have absolutely no grounds for believing that the factors that presented the problem for Heisenberg (i.e., the fact that on that scale radiation required for observation is a major disruptive force) are applicable. If there were interference by consciousness on the submicroscopic realm, still we would have no reason to conclude from science that the structures of reality above the level do not exist independent of our consciousness. The predictability of at least certain individual events in the macroscopic realm involves certainty, not a lower probability. That is, even if quantum events can affect other levels of organization within reality, still exact predictions of such events as eclipses have been possible. Billiard balls still behave like billiard balls regardless of the atoms involved. New forces are manifest on everyday levels, not a huge number of atomic events averaging themselves out. And for these levels the mechanistic approach may be appropriate, regardless of other levels.

Many other examples could be usefully discussed. To take a few: space-time may be a substantial reality on one level, but sound-waves still do not travel in the vaccuum of space. Time-reversal on subatomic levels might be occurring, but travel through history for more complex wholes cannot be justified be reference to this (entropy precludes this possibility as far as contemporary physics is concerned). Similarly, nonlocal effects occur only in very limited conditions in subatomic physics; they cannot be used to conclude that everything in the universe is affecting everything else regardless of space and time. Or consider the claim that "all is vibrating." Tables and chairs do not blink in and out of existence the way subatomic particles do. Thermodynamically irreversible processes dominate macroscopic objects, thereby preventing them from being adequately representable by wave-functions. So too the claim that "we are wave-phenomena" makes sense only when speaking of one level of reality's organization. The entire concept of "substance" may be a level-effect, an effect appearing only on the level of aggregates of molecules. If so, to speak of "solidity" as shown by quantum physics to be an illusion is erroneous. Gary Zukav misses this point when he says "Our experience tells us that the physical world is solid, real and independent of us. Quantum mechanics says, simply, that this is not so."[41] The concept of "solidity" refers to an everyday phenomenon, not to a lack of space through to the core of an object. Even if paper and a table are mostly space or are pure energy on one level, we still have no reason to think that one can pass through the other.

In short, it is hard to justify the claim that, because quantum physics is about all of reality (on one level), its claims are true on all levels. Such an approach denies the reality of structures on different levels of reality. It also is a metaphysical reduction of reality's multiplicity of levels to *one*. There is no simple microcosm/macrocosm parallel here; if there were, the linguistic problems in quantum physics would not occur. Oppenheimer was correct in saying that discoveries about the structure of atoms do not logically necessitate any philosophical conclusions about the world at large.[42] In a similar vein from Schrödinger: acausality, wave-mechanics, indeterminacy, and complementarity do not have as much connection with a philosophical view of the world as is currently supposed.[43] The philosophical implications of any scientific theories—constructs for understanding particular aspects of the world in a certain way—are not well-defined. Applying any theory to the everyday world cannot occur in a simple manner; a metaphysical expansion of the theory to cover all levels is necessary. Scientific theories may provide suggestions, but they are not themselves grounds for acceptance.

These problems have significance for the issue of the possible convergence of scientific and mystical claims. Mystical claims are related to meditative and other unusual experiences, but the interconnections that mystics allege occur on the *everyday* level of the world. How can they relate to scientific claims about the various submicroscopic or cosmological levels only? The best convergence seems to be that between Avataṁsaka Buddhism and bootstrap theory with regard to mutual interaction. A large portion of their outlines converge, but the difference in subject matter (i.e., what aspects of reality are under study) remains essential. To say that mystics see on a macroscopic scale what scientists find in their theories—or, in the words of Alan Watts, that the Buddhist and scientific views of the world as a field of related functions are the same except that the Buddhist vision is experiential rather than theoretical[44]—is misleading. The major difference between the highly refined theoretical framework designed for special experiments and the metaphysics of the everyday realm is being ignored. In addition, if what physicists have discovered can occur only on submicroscopic levels, one might ask how this could be what mystics have in mind. Mystics do not have increased powers of vision enabling them to see quantum-level phenomena. Nor do they see everyday objects blinking in and out of existence or see these objects as consisting largely of empty space. Rather their attack on substantiality as mentioned earlier is only the everyday notion of impermanence applied more thoroughly and experienced through meditative concentration. Solidity is not a special topic to mystics. Nor in their view do we have to reach another physical level to find something more real whose interconnections we then must apply to the everyday realm. Only two levels are important to the mystical point of view: the dreamer and the dream; no level within the dream is more real than another.[45]

The differences in experiences and levels of reality covered mean that science and mysticism will not converge or conflict in substance rather than in abstract categories. Mystics claim impermanence for the realm of experience and for all of reality. As mentioned earlier, finding any permanence on any subatomic or cosmological levels may prove a problem for the Buddhist no-self doctrine but not for the claim that constructed entities are impermanent. Would finding a total lack of permanence in the subatomic realm *confirm* mystical claims? No, since mystical claims are about all of reality, these claims are wider than the accumulation of the scientific claims. Mystical experiences of wholeness and oneness are not going to be scientific evidence for a theory about quantum physics, and the reverse is just as true. If science is relevant to nature-mystical metaphysics, then it may contribute to its falsification (by finding permanence) but not to its confirmation, since more is being claimed here than in science. If scientific claims about the everyday realm do not prove or disprove claims about the submicroscopic realm but are simply irrelevant, then this relationship between mysticism and science should not appear odd in comparison. Submicroscopic levels are not "consistent" with mysticism in a way that macroscopic levels are not (since mysticism is about the macroscopic too). Bootstrap theory has not "demonstrated the reality in mathematical terms"[46] of mystical connections—or even that of any other scientific interconnection. And applying it by analogy to mysticism would remove its scientific warranty. In light of this, even the more modest claim that "modern physics lends some support" to the "mystical vision"[47] cannot be accepted. As was concluded under the previous topics, scientific theories about subatomic realms cannot justify mystical mindfulness in principle either; if events in these realms permit either ordinary or mystical experiences to occur on the everyday level, they cannot justify only one as cognitively superior to the other.

(3) Specific Thinkers

To go into detail on the work of three popular thinkers discussing Western science and Eastern thought is worthwhile. Other authors could be discussed, but these present a variety of positions and, unfortunately, also of errors.

F. S. C. Northrop

Northrop does not write on the relation of specific theories but on the relations between the "aesthetic" (i.e., experiential) and the "theoretical" in generalized "Eastern" and "Western" cultures. According to him, all Asia focuses upon the immediately apprehended, for example, "blue" as experienced as opposed to "blue" as a theoretical construct concerning wavelengths. The realm of experience (the aesthetic continuum) is either "differentiated by specific sense qualities conveyed by different senses"[48] or undifferentiated. The latter is the former simply free of concepts of differentiation. The timeless, immortal, all-embracing, undifferentiated experiential continuum is referred to in Asia by such terms as "Brahman," "void," "*tao*,", "*nirvāṇa*," and "*jên*" (p. 382).[49] All sense-experiences are transitory differentiations of it. The undifferentiated whole is also a concrete particular immediately present in any sense-experience, but it is experienced as such only in yogic experiences.[50] It is also ineffable, since any language would make differentiations and lose the immediacy of apprehension. Even the experiential knower is one with the experiential object: introspected pain and sensing the blueness of the "aesthetic sky" are the same with regard to the identity of the experiencer and the experienced.

To contrast with this Asian approach:

> What the West discovered is the existence of a factor in the nature of things, not immediately apprehended, which only theory can designate and which only indirect verification through its deductive consequences can confirm or deny. (P. 305)

The experiential component is reduced to mere appearance both in science and in common sense. Even Western religion identifies the divine with the theoretical component.[51] By means of hypotheses and indirect modes of verification, we are able to learn of structures, orders, and entities that have a character different from those of the immediately sensed qualities in the immediately apprehended aesthetic continuum. Hypotheses always remain tentative, since the hypotheses assert more than what any amount of experience entails. Material and mental substances are posited. And in place of the experiential continuum, there is the theoretical continuum: the mathematically defined space-time continuum inferred from the aesthetic continuum.[52]

Concepts (terms to which meanings have been assigned)[53] differ according to whether they are experiential or theoretical. The former have their meaning assigned by intuition (immediate experience) and the latter by postulation. They should not be confused: aesthetic concepts are names for particulars while theoretical concepts are syntactically tied to some deductively formulated theory. The error of postivism (i.e., the position that there are only concepts by intuition) is that we cannot get mathematical physics from sense-experience—there is an unbridgeable gap between the two types of concepts. All Asian concepts are aesthetic; for example, "mind" in Zen and Taoism is a concept by intuition and is not one whit different from the immediately sensed body or nature.[54] Scientific and commonsensical concepts have reference to unseen factors, which can be known only by inference (pp. 378–379). Even if the *terms* denoting them are the same, there is a difference between sensed and mathematical "space," or between "blue" as a sensed color and a number of a wave-length in electromagnetic theory.

In Asia philosophical systems only direct attention to the immediately aesthetic factor that for these systems alone is *real* (p. 364). The undifferentiated experiential continuum is more basic than the differentiations, but both experiential factors are real for the East. The theoretical component is the illusion. In the West, Hume accepted only the aesthetic differentiations as real, but scientists such as Newton and Galileo overemphasized the theoretical: objects in "true, real and mathematical" space and time are all that is real. But, Northrop argues, both the experiential components (the undifferentiated continuum and the differentiations) and the theoretical component must be accepted as real: the three-dimensional theoretical object and the two-dimensional experiential components are both real, irreducible, and equally primary. The immediately experienced self and the persisting physical-chemical-biological body known by indirect verification are both real. The two approaches (experiential and theoretical) need each other: they are not the same, but are complementary (in the ordinary sense) (p. 454). For Northrop, the correct relations between the empirical factor and the theoretical factor is specified by the method of Western science: correspondences or epistemic correlations between the empirical and the theoretical, not a three-term relation of a subject, object, and appearance (p. 442).

Many objections of varying importance can be raised to Northrop's position. Certainly the idea of the unity of Eastern cultures is doubtful—either in a specific mystical/philosophical system or in the claim that *all* Asians are mystics. His claim that no Eastern thought in an "uncorrupted form" is theistic (p. 401) falters in the light of some Upaniṣadic thought alone.

Most serious are problems with his ideas related to concept and experience. All Asian concepts, he says, are concepts by intuition, which gain their *complete* meaning from the experiential component.[55] In the chapter on knowledge, it was argued that concepts such as "*kamma*" are no less theoretical than "gravity." Each involves some reference to experience and some to theory. Regarding the mystical, which Northrop equates with the undifferentiated aesthetic continuum, if all Asian beliefs are given immediately by experience alone, why are beliefs not identical and static? To argue that Advaita and Sāṁkhya-Yoga, to cite only two traditions, are the same is difficult. How Śaṁkara's "Brahman" differs from Western concepts such as Plotinus's "One" and Eckhart's "Godhead" with regard to at least this epistemological issue is hard to see. He identifies the theoretical component as the illusion that *some* Eastern traditions discuss. But the idea that for Asians the aesthetic differentiations are real too (p. 341) goes against some traditions discussing illusions: for Advaita, for example, any differentiation (whether experiential or theoretical) taken to be real is nescience.

His claims concerning Eastern empiricism fall into the same problem. It was argued in Part II that mystical enlightened states do not fit the ideal of empiricism: differentiations based upon theories are involved in these experiences as with ordinary sense-experiences. If this is so, then the claim "Far Eastern religion is positivistic, empirical, and, hence, scientifically veridical religion"[56] is wrong. In the extreme nature-mystical experience (which cannot be sustained throughout an enlightened state), the empiricist ideal is satisfied: sensory-experience free of conceptual structuring occurs. In that limited instance, Northrop is correct in calling Asians "relentless empiricists." Still, this is only one type of nature-mystical experience. Normally those experiences, like ordinary ones, involve concepts directing attention to aspects of the flux of reality. A nature-mystical experience differs by degree from ordinary experience in the degree of control that concepts wield over the sensations, even though Northrop would group them together as differentiating experiences. For him, *all* sensory experience ultimately is mystical. But what also fails is his neat distinction of concept-free experience and postulates indirectly verified by those experiences: concepts play a part in the experiences themselves. Science and mysticism have different concerns; but science and nature-mysticism cannot be contrasted absolutely on the role of concept-guided experiences, and this fatally undercuts his program.

Also there is the depth-mystical experience, which differs from the nature-mystical and more mundane experiences in being totally nonsensory while still allegedly a conscious insight into the nature of the world. This experience does not appear to have a place in Northrop's system. He says we can never achieve an experience of the undifferentiated aesthetic continuum since some differentiation is present in any experience (p. 335). But in sensing any specific aesthetic differentiation we immediately apprehend the entire continuum (p. 336). The undifferentiated continuum can be known by *abstracting* it from the local limited differentiations; this involves the "scientific" procedure of yoga (p. 342). This seems to make yoga not a mystical experience but the recognition of an abstract universal category. At best it would be the extreme nature-mystical experience of the concrete background free of any differentiations. The undifferentiated experiential continuum remains extended and differentiated even if it is not an object and the differentiations are not experienced. The depth-unity is not ever experienced as part of a continuum but as absolute identity. He is going against at least Advaita in claiming that Asians consider all things "made in part out of" the undifferentiated aesthetic continuum (p. 396). The emphasis upon experience is appropriate, but all things are identically the mystical, not made from it or having any other source of reality.

Fritjof Capra

Northrop deals with broad philosophical issues. The other two thinkers to be examined here make more specific comparisons between Western science and Asian thought. Fritjof Capra is a physicist who appears to have undergone a nature-mystical experience, an experience in which he felt atoms in the sand in motion and which he took to be Śiva's dance (p. 11).[57] Unfortunately, neither fact necessarily insures that he is aware of the methodological and philosophical issues of comparing science and mysticism. He does realize some fundamental points: that observations in both enterprises are theory-laden and that both Eastern images and physical theories are the creations of our minds (p. 44). Significantly, he also realizes that mysticism goes much farther in denying the difference between the observer and the object—there is not just a connection of things as in physics (pp. 141–42, assuming this is a correct understanding of physics). Some of his remarks on Asian thought are excellent, for example, his presentation of the problem of suffering for Buddhism (p. 95). Others are misleading. His interpretation of *māyā* as the illusion of taking concepts for reality, of confusing the map with the territory (p. 88) misses the central point of the relation of *māyā* to Brahman: in Śaṁkara's framework, no concepts, no matter how we treat them, reveal anything of the ultimate nature of reality. Unfortunately too, Capra generalizes in terms of "mysticism" and "Eastern thought" in speaking of one mystical holistic world-view, ignoring traditions (e.g., Advaita and Sāṁkhya-Yoga) that go against such a notion. For him, Eastern philosophies partially contradict each other, but the basic features of their world-view are the same.[58]

His basic position is that science and mysticism are complementary approaches that converge on the same world-view. Each approach is a distinct and irreducible way of knowing: the physicist experiences the world through the extreme specialization of the rational mind (which deals with distinctions and mental constructs) and mystics experience the world through the extreme specialization of the intuitive mind (p. 306). He does concede, though, that both intuitive and rational knowledge do occur in both mysticism and science (p. 30). Scientists are interested in explaining things, while the Eastern sages are concerned with a direct nonintellectual experience of the unity of all things achieved through stilling the rational mind (p. 230). Each approach inquires into the essential nature of things: one into consciousness, the other into the reality behind everyday life (p. 304). And each reaches a holistic world-view: the world is basically one and the parts can be properly understood only in the web of relations to each other and the unified whole. Mystics see the fluid, everchanging nature of phenomena in the everyday world while scientists see the same thing in quantum physics and relativity. Capra therefore speaks of consistency and harmony between physics and Eastern mysticism (p. 303). Contemporary sciences "echoes" or "rediscovers" ancient ideas (pp. 154, 270). Thus mystical insights are confirmed by physics (pp. 114, 161, 223). Physics, the "hardest" science, "has now led us to mysticism."[59]

Capra, unlike most writers in this area, does realize that the mystical claims involve the macroscopic level and that the physicists' claims involve subatomic particles or relativistic sizes (pp. 142, 204). He rightly says: "The fact that the Mahāyāna theory of interpenetration involves macroscopic objects, whereas the bootstrap model concerns subatomic particles, does not affect the similarity of intuition."[60] But he does not realize the significance of this divergence. He thinks mystical thought "provides a consistent and relevant philosophical background to the theories of contemporary science, a conception of the world in which scientific discoveries can be in perfect harmony with spiritual and religious beliefs."[61] This ignores the problem of differing levels; for example, from the scientific point of view we can speak of separate objects on the macroscopic scale (p. 310), precisely what Buddhist and other mystics deny. Although bootstrap theory is currently out of fashion among most physicists today,[62] Capra may be correct in asserting that if this theory is successful for hadron physics, physicists will try to extend it to the other interactions, perhaps to macroscopic space-time and consciousness.[63] Applying the idea to another scientific realm however, or working out a metaphysics with these ideas as the root metaphor is not something deducible from the theories themselves. When Chew speaks of a bootstrap philosophy of nature or of carrying bootstrap to its logical extreme of including consciousness (pp. 286, 300), he is making a complicated maneuver appear simple: he is removing the tentativeness and scope of the specific scientific theory and so enters another set of problems (e.g., how to check empirically the new claims). Similarly when Capra says that "the world view implied by modern physics is inconsistent with our present society" (p. 307),[64] he is overlooking the problem of the applicability of concepts from one realm to another and the problem of the relation between scientific claims and ethics. For the former problem, he himself claims that the holistic world-view has little value for science and technology on the human scale (p. 304). Toulmin's discussion of the relation between ethics and evolution is directly applicable to the latter problem here.[65]

On other specific comparisons he errs too. For example, mystics have nothing comparable to a conception of unified space-time (pp. 150, 164f.),[66] nor does the mystical "now" have anything to do with relativity (p. 179), as was discussed in Part II. His reasoning is often of this sort: because science and mysticism each have difficulty with language, they are talking about the same thing.[67] In addition, he relies too heavily upon abstract commonality between scientific and mystical claims of the lack of permanence among objects. Occasionally, he does glimpse more accurately the relation of science and mysticism. Thus he concedes that modern physics because of its framework cannot go as far as mysticism into the unity of things (p. 142). Or, while insisting that the intuitions of science and mysticism are closely parallel, still the underlying reality

> of the Eastern mystic cannot be identified with the quantum field of the physicist because it is seen as the essence of *all* phenomena in this world and, consequently, is beyond all concepts and ideas. The quantum field, on the other hand, is a well-defined concept which only accounts for some of the physical phenomena. (P. 211)

His assertion that mystics deal with the root and scientists with the branches (pp. 306–7) should point to the differences in their interests and caution us as to other possible differences.

Capra does not want a synthesis of mysticism and science: we need both (pp. 306–7). But he thinks that physics is a way (Chinese, *tao*) to spiritual knowledge and self-realization (p. 25).[68] It is not that mysticism is liberation from time and that "in a way" the same may be said of relativistic physics (p. 187). Rather, there is a spiritual dimension to Eastern thought, and current science is becoming a science with spiritual dimensions. (The discussion in chapter 3 should raise doubts as to whether science is becoming progressively more mystical in its nature, though.)

Some things would suggest the differences; for example, physicists are satisfied with approximate knowledge while mystics speak of attaining final knowledge (p. 290). In addition, the Advaita Vedānta concept of "Brahman" indicates a major divergence from what science deals with. From Capra:

> The Eastern world-view is intrinsically dynamic and contains time and change as essential features. The cosmos is seen as one inseparable reality—forever in motion, alive, organic, spiritual and material at the same time. . . . The two basic theories of modern physics . . . exhibit all the main features of the Eastern world-view.[69]

Brahman becomes merely a unified field out of which everything arises (p. 211). According to Capra, the highest aim of Eastern thought is to become aware of the unity and mutual interrelationship of all things, to transcend the notion of an isolated, individual self, and to identify oneself with the ultimate reality.[70] In sum, the most important characteristic of the "Eastern world view" for Capra is the awareness of the unity and interconnectedness of all things and events, and the experience of all phenomena as manifestations of a basic oneness (p. 130). This in no way accords with Advaita which emphasizes the unchanging, timeless, still center rather than a surface-flux. Only the former is real for Advaita: Brahman is not an extended, structured field but a dimensionless awareness. Any account of Eastern thought that ignores this must remain a truncated one.

Gary Zukav

Gary Zukav, a layman making brief comparisons with Eastern thought while surveying contemporary physics, makes some errors worth noting to show the depth of misunderstanding so often involved in such comparisons. Some of his errors concerning Eastern thought are minor. For example, "Buddhist practice is called *Tantra*" (p. 330)[71] is simply wrong: "Tantra" refers to particular Indo-Tibetan traditions, and "practice" (actions and meditation) has a central role in all Buddhism. Zukav's attribution of whatever he values in Buddhism to Tantric Buddhism (e.g., p. 217), even things that *all* Buddhists would endorse, also shows a lack of understanding. Other errors are more costly: there is nothing in Buddhism to warrant considering it a form of "superdeterminism" (p. 318). "Free will" is necessary if we are to be able to choose a course of action leading to enlightenment; control and nonrandom change, but not fatalism, are necessary. So too when Zukav says that "true" scientists and mystics do not seek truth and reality but are only interested in playing their respective "games" (p. 111), certainly most mystics and scientists would be surprised.

His comparisons are disconcerting at best. Here are three examples. According to him, Buddhists treat reality as consisting of "virtual entities," that is, similar to those subatomic particles which appear briefly but do not remain as more stable entities having a degree of independent existence.[72] There are no "bare" particles free of interactions at this level; instead every "real" particle has a cloud of "virtual" particles surrounding it. To see Buddhist impermanence and dependence in this way can only distort: the impermanent aggregates may remain intact, although without substance, for a long period of time (appearing brief only from the point of view of an infinite number of cycles). The "real" quantum entities are just as impermanent as the virtual quantum particles from the Buddhist point of view. In addition, there is nothing comparable in Buddhism to the virtual particle interactions: virtual particles apparently arise randomly, not in the orderly process as with Buddhist dependence. No Buddhist notion of dependent and impermanent entities parallels the idea of a cloud of entities popping in and out of existence or of temporary particles that decay into more stable ones. Nor is the distinction between virtual and "real" particles similar to the distinction in Buddhism (and many other metaphysical systems) between reality as it actually is and as it usually seems to be (contra p. 253).

Zukav, like many others, also sees paradoxes in physics as Zen *koans*. *Koans* are linguistic devices used in conjunction with other meditative techniques—a type of paradox voluntarily introduced into mystical training.[70] When a Zen master answers the question "How about the ancient mirror not yet polished?" by calmly saying "Han-yang is not very far from here" or on other occasions exclaims "Listen to the sound of one hand clapping!" the Zen disciple is being given a verbal, intellectual beating very much like the more physical kind (which is also employed where it seems appropriate). The attempt seems to be in a very startling way to direct the disciple's attention away from assertive claims and ideas to which the intellect may become attached (thereby hindering the religious quest) and to deny that any answer "fits the case." The disciples need to learn that they have not attained the truth if they feel that the questions posed are appropriate to reality and answerable. The *koans* are very much like the questions of children who do not understand the construction of reality ("Why is this stick bigger than that one?" "Why is New York an island?" "What does the wind do when it is not blowing?"). When the students realize that any "serious" reply to the questions or commands presuppose a world of distinct entities (rather than how the world is actually composed), they are ready for a nonverbal experience of reality. After this enlightenment, they can use the old language again unaltered; only the tendency to project concepts onto the world has ceased.

Zukav claims that "Picture a massless particle" is a scientific *koan* (p. 224). This may be a poor example, since scientists have been able to work without visualization; they do not exhaust themselves trying to picture massless particles or any other such entities. But most important, *koans* represent a nonassertive approach to reality (i.e., no assertions about reality are involved), while science always remains assertive. Working through paradoxes (assuming they are not just superficial ones) may also be part of both science and mysticism, but their scope is different: "enlightenment" in physics is only to a new theory of limited scope, not to a new vision of all aspects of reality. We might also wonder how much attention is focused upon the paradoxes themselves in actually doing scientific research. Any breakthrough may sound mystical until a new conceptual framework evolves. In any case, science advances by means of new conceptual schemes, while in mystical enlightenment only the old conceptual system is seen differently in light of the old (timeless) knowledge newly internalized.

To deal with a final example, Zukav, without attempting a justification, treats consciousness as a form of energy. For him body-and-mind is just another form of the energy-mass identity (pp. 176–77). They are not analogous, but two statements of the same identity. This is very hard to accept. Energy is a measurable physical factor; awareness might be too, if we accept paranormal powers. But even if we accept that idea, still it is something else to argue that body and mind are two manifestations of the same thing, or, giving priority to mind as with energy, that the body comes from the mind as a temporary "form" and returns to the state of energy. To say that science expresses these two points in accepting "$E = mc^2$" is simply absurd. To assert further that in the East there never was much confusion about matter and energy because of this (p. 177) is also wrong: "energy" in any recognizable scientific sense and the relation of it to matter are not topics dealt with in any classical Asian religious tradition. Nor are mind and body related in that manner in any of these traditions. Zukav varies the example, but not the reasoning, when he finds that Bohrian complementarity "in human terms" means that the same person can be both good and evil (p. 220). Taking quantum physics as proof of claims in the everyday realm is as impossible as taking theories of our psychological nature to justify physicists' claims. So too, arguing that mindfulness received the validation of Western science via Minkowski's rigorous confirmation of space-time as an independent reality (p. 176) is a paradigmatic illustration of this error: mystical mindfulness involves being aware of only what is happening here and now; it has nothing to do with exotic theories of the relationship of space and time.

Conclusion

Examples from other sources could be analyzed. A line from an old Persian poem—"If you dissect the heart of an atom, you will behold a sun within"—may suggest parallels to the speculations that our universe is a black hole in another universe or is an electron in another universe (and that each electron in ours contains another universe). But none of the alleged convergences ever extend far beyond abstract parallels. Much is made of Buddhist impermanence, but its notions of dependence never correspond in any detail with scientific ideas. Avataṁsaka does correspond to bootstrap theory to this extent: both speak of mutual interdependence. This degree of convergence cannot be denied, but to go beyond this one aspect of this one Buddhist school and one scientific theory to claim that science and mysticism are converging in any substantial way is unwarranted. The total contexts of mysticism and science are so different that science cannot be taken to be any specific mystical tradition revived in a Western form. Depth-mysticism cannot be ignored nor can all nature-mysticism be made into interdependent holisms. The parallels amount to little more than the notion of impermanence present in nonmystical Western thought since Heraclitus. Finding impermanence on a submicroscopic level is "consistent" with mysticism but unnecessary; mystics find all the impermanence they need on the everyday level of reality, and this quantum physics cannot support. Science cannot validate either mystical experiences or any particular mystical belief-system. And considering that such divergent interests, methods, and cultural milieus are involved, we should not really expect more than abstract parallels to obtain. Mystical interconnections and structures appear commonsensical compared to subatomic and cosmological scientific ones, and they appear bizarre in the thoroughness and grandeur of scale of their application.

The significance of the parallels is more than noting that the alphabet of Sanskrit (i.e., *devanāgarī*) may look something like mathematical equations to the uninitiated or that *"tao"* is pronounced "Dow" as in "Dow Chemical" or the "Dow Jones Average." But the type of reasoning employed by most of these writers in their enthusiasm is comparable to science-fiction writers going from scientific theories of anti-particles to positing anti-universes populated by anti-Captain Kirks and anti-Mr. Spocks. Science will not become mystical in the sense of involving mystical experiences simply by the adaptation (through analogy or otherwise) of interesting elements of the metaphysical framework of a mystical system, any more than mystics will become scientific by using the idea that gravity is holding us to the earth as a meditative tool.

PART IV

A Reconciliation of Science and Mysticism

10
A Reconciliation of Science and Mysticism

This final chapter will attempt to outline a reconciliation of mysticism and science based on the positions taken in the previous chapters. In this reconciliation, cognitive value will be given to both science and mysticism. Mysticism will be assigned authority in matters of "being," and science in matters of structures. Thus both the depth-mystical and the diversified structures will be considered real. In this way, this reconciliation is a dualism, although there is only one ontological source, somewhat similar to that of the Neo-Confucian Chu Hsi, who accepted a multiplicity of "principles" *(li)* and a unity of material force *(ch'i).*

This is not an integration of science and mysticism in the sense that mystical claims should affect scientific theories or vice versa. Nor is it an attempt to reduce either enterprise to the other with respect to methods. It is, however, a reconciliation. That is, it finds value in each and places both in one overarching conceptual scheme. Both enterprises are accepted as separate and necessary for a fuller understanding of reality. (It may also be that both enterprises are necessary for living the fullest life, but this claim is not identical to, or deducible from, the epistemological one.)

Such a reconciliation is also a work of metaphysics, in that a coherent conceptual scheme encompassing two separate experience-related enterprises is being advanced. It is not, however, a speculative metaphysical system that systematically places all experiences and all enterprises into one scheme. Nor will answers be advanced concerning all important philosophical questions raised by the undertaking. Only the issue of science and mysticism as enterprises is addressed. For this, no specific scientific theory or full mystical system needs to be defended. The mystical oneness is given an ontological function as the ground of reality, but the exact relation between this ground and all other elements of a world view (e.g., the exact relation between this ground and individual selves) or the role of the mystical in a way of life (e.g., its relation to the problem of suffering) and any further specification of the nature of the mystical will not be dealt with.

The Problem of Justification

One preliminary philosophical issue—justifying any position on science and mysticism—must be discussed in some detail, however. A reconciliation is certainly not the most obvious or easiest position regarding science and mysticism—rejecting one enterprise, and thus any reconciliation is easier. Nor is only one reconciliation possible. How can one justify the reconciliation chosen here? A complete justification of this or any position does not appear possible. It is a matter of final beliefs, a matter in which unanimity does not seem achievable.[1] Each of several possible positions relating science and mysticism depends upon values and choices that remain assumptions not admitting of complete justification. There are no timeless and cultureless grounds for adjudicating between possible positions. Since we cannot free ourselves from all particular points of view, acceptability may ultimately depend upon grounds internal to the position being defended.

To whom then do we have to justify our beliefs? Is each of us the final court of appeal for ourselves? Is the choice between positions relating science and mysticism *rational?* Part of being rational is accepting only those beliefs for which we have reasons. But the circularity returns: divergence on what is fundamentally real affects what is accepted as known and "reasonable." It determines what is accepted as a "fact" and thus cannot be settled by the "facts." Holding one's beliefs open to criticism also is necessary to being rational, but it is not sufficient for deciding which position to accept. Likewise, some consideration of what is *prima facie* counterevidence is necessary. (Here, arguments against the cognitive significance of both mysticism and science will have to be noted.) Arguments between holders of different positions do appear possible, but in matters of "ultimate" importance, a choice must be made; we cannot remain neutral nor will the issue be decided empirically.[2] This does not need to lead to the view that we necessarily know nothing, or that all claims are equally legitimate (i.e., relativism). Still, it does mean that we should treat all our beliefs tentatively in light of what in the final analysis is the lack of a logically certain "ground of belief."[3] We accept such beliefs with a lack of true evidence. And thus, as Heisenberg says in another context, the most important decisions in life must always contain an inevitable element of irrationality (i.e., of pushing arguments aside and deciding).[4]

The choices made here regarding the significance of mysticism and science are that each enterprise increases our knowledge of reality. If science needs no nonscientific justification, then neither does mysticism need a nonmystical one. But counterpositions can be stated too—that is, that either scientists or mystics or both misinterpreted the significance of their experiences, and in fact only one enterprise or neither one provides an insight into reality. Different constructions of the world would then result.

Mystical insights have often been challenged. They have been characterized as delusions resulting from a malfunctioning brain or given some other naturalistic reduction. The loss of a sense of a self or of time is explained as relating only to the experience, not to some reality. Such experiences may be self-induced illusions, which seem so powerful to the experiencers that they cannot help considering them true insights. This position can be supported by claiming that mystics begin with a strange idea and go through mind-stressing exercises until they become convinced that their idea is correct. The prescriptive nature of mystical perception (e.g., viewing one's own actions and body as alien) also raises doubts as to whether the mystics see things "as they really are." Certainly the Buddhist claim that the world is suffering and is to be rejected, and the Jaina monks who give up their lives passively, lend credence (in some circles) to the idea that all mysticism is delusory. Arguing that pain is an illusion has the same result. A mystic's reply is that the value attached to survival is delusory in light of the true nature of reality—as revealed in mysticism.

Doubts can also be raised concerning the foundations of the scientific enterprise. Science too has assumptions and is prescriptive. Scientists look at the world with definite purposes and methodological limitations. There is the danger, as Dōgen says, of man's disposing himself toward the world and construing his dispositions as the world. Reality may be scientifically understandable only because we force it to be so—we would not permit it not to be.[5] Is there any guarantee that science tells us something of reality and is not just our own prejudices reflected back? The regularities and intricacies of solar mythology or pyramidology would then be the paradigms of science.

The objectification of reality in science and common sense may be one instance of such a distortion. The scientific attitude involves a definite restriction on the mind: it directs the mind to a world of posited entities. It also divides the world up into subject and nonsubject. Is Kierkegaard correct in asking if our activity as "objective" observers of nature will weaken our strength as human beings? Does such a quest alienate us from the rest of the world and ourselves? Does it remove us from the fundamental nature of reality? To take a different angle, can scientists claim to *comprehend* reality by finding some regularities in a sea mostly of chaos? Is it not rather audacious to claim to comprehend even some aspect of the whole universe from a limited amount of experience by one species located on only one small planet? Or how do we know that our experiments are not comparable to "distorted room" photographs whose effect results from the camera lens alone? Jacob Needleman has an analogy that may be very close to the truth: scientific theories "work" like someone ignorant of guns using a gun for a hammer; since experience shows we are correct, we may never begin to ask the right questions to discover the proper nature of the gun; we may even improve the gun as a hammer by removing the bullets.[6] He concludes that successes might only be signs of the sort of questions we are actually asking of reality rather than answers to questions more appropriate to reality (i.e., questions that reveal more of the nature of reality).

From all of this we can at least conclude that neither science nor mysticism provide self-evident insights into the actual nature of reality. Science's position can be questioned, no matter how much authority science has in our culture because of historical contingencies, without the charge of irrationality being obviously justified. Ultimately, a judgment is involved—acceptance of science cannot be asserted dogmatically. Again, there are limitations to rational discussions. At best, all that can be done is to state one's assumptions. The assumption for the conciliatory position advocated here is that both mysticism and science give cognitive insights into the nature of reality.

Statement of the Reconciliation

The reconciliation that suggests itself from the positions taken in this project can be simply stated: mysticism, with its emphasis upon the experienced "being," will decide issues of the fundamental that-ness of reality, and science, with its emphasis upon regularities within the realm of change, will decide issues concerning the how-ness of reality. Thereby the strengths of each approach are accepted without rejecting the other. Only metaphysics that reject one approach or the other are rejected. This solution is not anti-scientific: scientific concerns do hit an aspect of what is fundamentally real—the structures producing order in the realm of change. But science does not provide the exclusive way of knowing reality. Scientists do not deal with abstractions unattached to the world we live in, but with relations; mystics do not deal with uniqueness, but with being. That is, we live in a world surveyed by both mysticism and science, with different types of constants (structures and being) being sought in the different approaches.

Both structures and "being" (the ontological ground or dimension of reality) are real: the former component is responsible for order in change in the realm of becoming; the latter is the "power" that supplies what is ordered. We normally associate reality with "being" or substance—just as the term is singular, so we assume that all reality is ultimately one. But this is just an assumption. Reality is *one* in its substance and *many* in its structures. There are not two ways of experiencing one reality, but two equally fundamental components of reality. Both the structure and the depth-substance are reality "as it really is." The beingness is experienced and the structures inferred (roughly as Northrop asserted). There are many levels of organization and thus many structures to reality. A structure is real if it is responsible for some relation to change among phenomena. The structure is just as real as what is structured; neither is the sum of reality. There may ultimately be only one source for both, but there is no reason to equate the utterly simple being with the complex structures since each is equally real and irreducible.

Thus there are two basic types of knowledge: knowledge of the order in diverse phenomena and knowledge of being. Both types are separate and necessary. Mystical experiences involve awareness of the being: depth-mystical experiences realize the identity in a nonsensory experience free of all differentiations; nature-mystical experiences are experiences in the realm of becoming of the common being underlying all phenomena. In sense-experience structured by concepts, the "pure" being is not sensed. Two fundamental errors are possible: in the matter of being, not realizing reality's identity; and regarding structures, not finding the actual fundamental structures. In each case we may erroneously accept surface-appearances by affirming a multiplicity of real entities or by mistaking our constructs for reality.

The scientific approach reveals nothing of being-ness, and the depth-mystical nothing of structures. Each type of claim has its own context, and only within its context does either enterprise reveal how things really are. These contexts or domains are limited by the type of experience that is cognitively central and by the concern involved. Each approach is prescriptive to that extent and neither is exhaustive of the total picture of reality. Both make claims about reality but about different aspects of it. They can conflict over ontology or structure only if the context of the claims is forgotten. The claims are incommensurable and so can be fitted into a coherent world-view.

Scientific knowledge is "lower knowledge" only in the sense of dealing with something other than the ontological dimension of reality. But to accept one source of knowledge and reject the other is to live an illusion. Mysticism can become as reductive as totally ignoring mystical experiences if it is taken as complete knowledge of reality. To deny the reality of the structures in the realm of becoming is to live in only an abstraction of reality. Mystical knowledge is a counterbalance to the manipulative trend, but it is more fundamental only in ontological matters. Thus for "wisdom," both types of knowledge must be accepted. They are complements in a vague, everyday sense.

Nature-mysticism stands between depth-mysticism and science. The impermanence and change it deals with are real, as is the beingness. "Being" has two levels: the depth stillness and the extended whole of the nature-mystical. Ontologically, the world of being and the world of becoming are identical, not related as source and dependent reality. The unstructured, nonobjectifiable depth-reality is the "substance" of the realm of structures, both material and mental. If the extensionless and the extended are ontologically identical, one mystery is how (and why) the extended "emanates" from the extensionless. How can the changeless and timeless be identical to, or even the ground of, the changing and temporal? Add to this the mystery of how "being" is related to structure. Is the structure independent of "being" or is it intimately tied to the creation-process as in the Greek concept of "*logos*"? At least the realm of change is structured, and the nature-mystical supplies the only substance. But this being so, the nature-mystical cannot be an explanation of any change therein: it is the beingness of everything concerned and adds nothing to the order of change. "Being" cannot explain why there are some structures and not others. The ontological dimension is one of depth, adding nothing to the horizontal linking.

Mystical mindfulness in its pure form sees the realm of becoming in its dimension of beingness—the "suchness" of this realm—without regard to the dimension of the structure. Scientific knowledge, a sense of time, basic feelings and emotions, and perhaps even concern for the body and survival must be suspended for any mystical experiences to occur. But mystical ways of life involve an understanding-component that may conflict with other metaphysical systems and with science. The nature-mystical explanatory concepts such as *"karma"* are theoretical and not directly observable. A conflict or convergence with science is possible on the order of structures. But in addition to levels and types of structures, there is the ontological order ("being" and becoming), and here science and mysticism cannot be related. That is, the relatively "close" levels of the broadest cosmological schemes and the nature-mystical reality are "close" only in the sense that this scientific structure has the least detail and so appears closer to the nature-mystical whole, but the fundamental difference between inferred structure and beingness remains. Conflicts may arise over what in fact are the fundamentally real structures (i.e., which structures are effective and not reducible to other ones). If there is one basic structure responsible for all other structures, it is not any closer to beingness than the others, nor can the unique properties generated on each level above or below the fundamental one be denied.

While mysticism deals with the being of reality, science deals with the variety of reality's structures. The basic division is between a concern for being and a concern for structure. With regard to structure, there is no *one* way that "things really are." Each scientific approach explains phenomena by describing a physical or biological structure of all reality, but only that one level is involved. Thus physics is complete with regard to its domain (the physical structure of reality) but partial with regard to the totality of reality. Science deals with simplified conceptual systems devised for selected purposes. These systems are like maps to selected areas; but maps distort in systematic ways that must be understood if we are to understand maps properly. Scientific abstractions are not in themselves illusory, any more than discussing exclusively the colors presented in the flux of experience reduces that flux merely to the colors. Only through misreading the nature of abstractions (mistaking them for the sum of the reality involved) do they become obstacles to experience and understanding.

Science corrects sense-experience (e.g., the sun is not smaller than the earth), and mysticism corrects science ontologically. Mysticism does not abolish the diversity of structures, but sees objects in relation to beingness. Scientific knowledge relies upon conceptual diversity and does reveal something of reality (the structures). Mystical knowledge transcends it in the sense that transcending a mirage is seeing that no water is actually involved: an object's proper ontological status is thereby revealed, but the sensations remain the same. It corrects the vision of things and gives science its proper place.

Each particular item is unique, not in its "being" (all "being" is one) nor in its structures (science understands it by subsuming it under its general structures), but in the unique interaction of the ultimately real structures. And in the context of making claims about structures and their interaction, scientific and everyday claims can be completely accurate. For example, the claim that "The New York Mets won the World Series in 1969" would not be true in the matter of being-ness, since events and time are unrelated to the matter of "being." To hold the view that there is a multiplicity of ontologically real individual beings is an error. But to deny that the claim is accurate in its context as no more than a surface-claim would also be a mistake. We are asking a specific question and not getting a false answer, unless we take the surface-multiplicity as the final ontological reality. Mysticism is necessary if we are not to "course in signs" as the Prajñāpāramitā Buddhists would say, that is, if we are not to take the sensations as indicative of a multiplicity of entities rather than perceive correctly in light of the ontological status of entities.

Knowledge-claims concerning ontological matters are also tied to one context and become deceptive outside that context. To say "all is one" (because of the one beingness involved) is clearly false in the context of structures. But the depth-mystical claims are the "ultimate truth" with respect to the fundamental ontological nature of the world. That is, they take all relevant factors into consideration for a final verdict in this one matter: mysticism here cannot be denied by anything drawn from surface-experiences on this matter. And since claims about ontology cannot refute or support statements about structure, the mystical and scientific claims cannot conflict.

Conventional claims can be used without projecting the terms ontologically (which would create a world of multiple distinct entities). Thus no error with regard to ontology is necessary. The language involved in the claims states something accurate about reality when the claims are true. This is so for both types of knowledge: there are "conventional" and "ultimate" claims in the context both of structures and of being. Language is drawn from the realm of structure and change. But if the mirror-theory of language is rejected, there is no reason to assert that, because language is part of one realm, it cannot state something about the other.

Such a solution may not be correct or generally acceptable. It will depend upon how mysticism and science in general evolve, and more particularly upon how metaphysical judgments of each enterprise evolve. But this position does show that, while many mysteries still remain, mysticism and science can be reconciled if neither is taken to be the only source of knowledge about the world.

Appendix

Concerning the Possible Philosophical Significance of Scientific Studies of Mysticism

A topic tangentially related to the issues of this project is whether natural scientific studies *of* mysticism (i.e., various physiological studies of mystics and meditators) have any bearing on the possible truth of mystical knowledge-claims.[1] Popular works on the physiology of various types of meditation often mention what their authors see as the philosophical implications of the study. The usual conclusion is that the study provides "a scientific *validation* of age-old wisdom."[2] Can such scientific studies in fact provide that?

Physiology and States of Consciousness

To determine whether there are any grounds even for maintaining that there is a scientific study of mysticism, let alone in what sense (if any) scientists can be said to verify mystical claims, three subjects of study must be distinguished: the body, states of consciousness, and insight-experiences. Ultimately, the issue at hand revolves around the relevance of the study of the first two subjects for mystical claims connected to the third.

For this, no attempt to define "consciousness" nor any metaphysical theory positing a causal relationship between "mind" and "body" is necessary. Only two simpler ideas need to be understood. First, within the realm of common sense, we distinguish consciousness ("mind") from the body ("matter") and feel that each can effect changes in the other. Familiar instances of bodily changes producing changes in consciousness are illness and pain-producing injuries. Examples of the reverse relation: "willing" one's arm to move, and fear producing perspiration and an increase in the heartbeat rate. Physiological studies of yogins have also documented conscious control over the "involuntary" nervous system, for example, control of body temperature and dramatically slowing the heartbeat rate.

Second, contemporary physiology finds it feasible to *correlate* in a one-to-one manner conscious states with neurological and biochemical states of the brain (or changes in these states), and with physical states of other parts of the body. For a true correlation, changes in states of consciousness (organized states in which awareness is structured for different activities) must accompany changes in physiological states (induced by ingesting drugs, electrical stimulation of areas of the brain, or whatever) and vice versa. By such correlations various states of consciousness can be studied through observable phenomena without necessarily delving into metaphysical explanations as to how or why conscious states are so correlated with physiological ones, for example, whether adrenalin flows because of anger, or vice versa, or whether the two events are identical. Any such causal arrow will have to go beyond the observable physiological mechanisms within the scope of physiological theories.

The first important questions are: What is actually being studied in such physiological and neurological research? Is consciousness being studied scientifically? Two considerations lead to doubts.

First, the indirectness of these studies must be emphasized. The issue is not how electrochemical brain-activity "becomes" feelings and the conceptual manipulations of thoughts (or however consciousness and body are related), but rather how by their nature scientific studies are limited to that which can be produced for direct or indirect inspection by others. What cannot be so observed must be ignored. Consequently, in a very important sense *consciousness* is not being studied at all in examining electrochemical activity, despite its one-to-one correlation with observable states. Charles Tart, commenting on the legal risk that trying marijuana incurs, points out that the primary effects of marijuana-intoxication that scientists discover (slight increase in heartbeat rate, reddening of the eyes, and so on) are completely distinct from the effects for which users risk jail: only the "subjective" change in consciousness is of interest.[3] This subjective change is what is actually experienced. But the experience cannot be presented for observation. Indeed, all experience is "subjective"; no experience can be presented for public examination, and hence is not "objective" in that sense. And with these physiological studies, the subjective can be ignored.

In principle, the sciences can give various distinct accounts (physiological, neurological, etc.) of any process, such as raising one's arm or seeing an object. Even if an interactionism is the proper metaphysical relation of mind and body, still complete accounts of the physical stratum are possible, and so there is a physiological basis in which patterns *may* appear. If this is so, all parts of the body are law-abiding mechanisms understandable from various points of view. Each account would give a complete description of an occurrence in the sense of accounting for every phase. But, to state the obvious, accounts of consciousness in terms of recordable electrical impulses across synapses between neurons is comparable to analyzing *Hamlet* in terms of the motor-mechanisms of Shakespeare's handwriting: the account would be complete, but something of value—another level of significance (here, the intent guiding his writing)—would never be touched. By means of electrophysiology, Robert Ornstein says, "dreams can be researched objectively."[4] So also we could study the Senate meetings scientifically by a simple correlation: when the doors to the Senate chamber are closed, we know the Senate is in session; when the doors are open, we know it is not. In neither instance will much of value be disclosed about the phenomenon under study. The "subjective" elements far outweigh the measurable aspects in importance. But at least dreams may have an as-yet-unknown physiological significance. In meditation, on the other hand, the prominence of the subjective elements is much more marked. Elsewhere Ornstein himself, while criticizing the idea that physiology validates meditative claims, likens electroencephalographic studies of meditation to placing a heat-sensor over a computer in an attempt to discern its program.[5]

Second, that observable phenomena can be systematically correlated with conscious experiences has been assumed so far. In physics, the sensations of color, whether a person is color-blind or not, can be correlated with wavelengths of light. And in light of the effect of consciousness upon the body and vice versa, it would seem reasonable to suppose, as Herbert Benson has, that religious prayers have definable physiological effects on the body.[6] Yet difficulties in correlating changes in physiological events with alterations in consciousness are indicated by such facts as there being no simple correlation between physical injury and conscious pain in every instance: either may occur before the other, and even in the absence of the other. To give another instance of this problem directly related to physiological studies: "under conditions of extreme sleep deprivation, the EEG can indicate 'deep sleep' when the subject is awake, at least by all the usual standards—talking, responding to instructions, etc."[7] Thus two vastly different states (being awake and being asleep) produce, under certain conditions, identical readings measuring one physiological state. Additional difficulties in such one-to-one correlations are revealed in the pharmacological placebo effect: once a response has been learned, we can be given what we think is the drug (but which in fact is not) and the response will occur; conversely, we can ingest the active ingredient (without being aware of having done so) without any change in consciousness occurring.[8] This effect holds for some drugs, including some powerful "mind-expanding" ones.[9] In short, different states of consciousness can be produced by the same stimulus, and different stimuli can produce the same conscious state: the physiological effects here are neither necessary nor sufficient for changes to occur. Therefore, the influence of the mind (our expectations, knowledge, and so forth) in effecting changes for some states of consciousness is more important than any possible biochemical process that may be involved and be measurable.

For mystical experiences, many problems arise with regard to finding unique configurations of physiological data for correlation. One is that there are two fundamentally different types of mystical experiences—one involving the sensory and/or conceptual apparatus of the mind, and one not. The former group involves many different experiences, while the latter appears to involve only one. Electroencephalographic indices differ for concentrative versus mindfulness meditation. But there are different degrees of meditative achievement from novices to masters. Will the enlightened mystics differ in their brain-activity? Will there be a unique correlation for each experience?

In addition, there is the problem of abstracting a physiological *cause* (a sufficient, or necessary and sufficient, condition producing a change in experience) from all occurrences of mystical experiences. Some mystical experiences occur "spontaneously" (i.e., unexpectedly). Many traditions speak of the need for "grace," because no action will *produce* the experience. No practices or physiological changes seem to *force* enlightenment: the internalization of the correct mental framework (correct knowledge and emotional detachment) either occurs or does not. A "leap" at the end of the "path" is required, according to most mystical traditions. For example, in Theravāda Buddhism as discussed earlier, *nibbāna* is considered "unconstructed" because it is not the product of any action or of an accumulation of merit. Disrupting our ordinary views of the world is necessary for enlightenment, and meditation (or, for instance, drugs) provides a means for accomplishing this.[10] But nothing can guarantee that we will let go of our ordinary views. The actual enlightenment is a noncausal switch of world-views, whatever the nature of the state of consciousness at the time. Mystics speak of what are the necessary and sufficient conditions—selflessness or whatever—but finding physiological correlates for them may be difficult. For example, a disease may be sufficient to touch off a mystical experience *only if* certain conditions preceded this for the experiencer. Certainly not everyone who has had any given disease becomes an enlightened mystic. Nor are all serene or joyful persons mystics. Abstracting a necessary and sufficient element common to all these situations for a one-to-one correlation may prove difficult.[11]

Even something as ordinary as perception cannot be correlated with its physical stimulus, as Gestalt psychologists emphasize. This indicates that there are different experiences independent of the state of consciousness involved. And this raises the possibility that there are physiological correlates only for the most rudimentary level of sensation or emotion; only claims for such a level would be falsifiable in any fairly simple, straightforward manner by physiological research. Otherwise, the transformations of consciousness seem to be an independent factor.

Physiology and Mystical Knowledge-Claims

For the study of mystical experiences, the above problems raise the possibility that mystical states of consciousness may not be correlatable with physiological data; different types of mystical consciousness may produce the same physiological readings, and different physical states may be present in the same types of mystical enlightenment. Even whether such an experience transforms a person may depend upon the setting and the experiencer as much as any experience in itself. Still, it does seem reasonable to assume that certain physiological changes occur during meditation. But if these changes remain constant regardless of the "subjective" response, then no one-to-one correlation is established. The physiological states would be at best necessary conditions to the mystical ones, not sufficient or necessary and sufficient (one-to-one) causes. The physiological changes brought about by meditation or disruptive drugs may or may not induce religious experiences, depending upon the belief-framework of the person involved (among other factors). Thus elementary concentrative techniques such as Transcendental Meditation may produce, say, an increase in alpha brain-waves (indicating a decrease in information-processing), an increase remaining constant regardless of the state of consciousness involved. Benson, in fact, finds that there is a great variety of "subjective" (experiential) responses—including no change of consciousness at all—accompanying the same physiological changes engendered by his relaxation technique.[12] In other words, electroencephalographic and other profiles of a person may be the same when one is merely deeply relaxing or has actually experienced any of a variety of alterations of consciousness. And the science itself will be neutral as to whether the physiological changes are due to changes in consciousness or vice versa, and as to whether the effects upon the body are more important than the effects upon the mind. The former point corresponds to Ayer's point that if a one-to-one correlation of the brain and mind is ever established, we would have as much reason to say that the brain depends upon the mind as vice versa.[13] A wider explanatory system would determine which we deem more fundamental.

As discussed in chapters 2 and 5, it is not pleasurable experiences or paranormal powers over matter that dominate mystics' interest in destructuring and concentrative meditation, but alleged *insights* into the nature of the world and the self that are discovered. An insight such as "awakening to the true nature of the self" contains an experiential component and a belief-component, whether the latter is an active component shaping the experience itself or is a more or less conscious interpretation applied after an experience that is totally free of all conceptual structure.

This belief-component constitutes the philosophically relevant "subjective" aspect of the insight. However, it is just this component that makes correlations of mystical insights even with states of consciousness difficult. In an enlightenment-experience, internalizing the previously studied conceptual framework (i.e., seeing that it is correct and that it applies thoroughly to all components of reality) may produce a change in dispositions resulting in a new state of consciousness. Or, in the enlightenment-experience and subsequent state, it may only be the knowledge that changes, in which case the state of consciousness remains the same: only an experiential change of a Gestalt-like switch in perceptions would occur (i.e., sensory stimuli would be reshaped by the new knowledge), and not a change in how our awareness is organized. It should be remembered that the Buddha's enlightenment occurred only outside of the lucid trances *(jhānā),* although his mind was "concentrated." In Indian mystical systems, it is realized that concentration *(samādhi)* occurs in one degree or another in all states of consciousness and in all experiences and that it cannot in itself insure enlightenment. As previously mentioned, nothing can force the internalization of the correct cognitive and dispositional framework. Those traditions, such as Advaita, which rely upon the deeper mystical experience (where the mind is completely emptied of all normal content, sensory and conceptual) as fundamental tend to identify one state of consciousness with the experience itself, and this with the insight. But the conceptual scheme for understanding the experiences after the enlightenment-experience reveals the conceptual dimension to the insight. Śaṁkara uses the Vedas to understand properly the mystical experience. Thus mystical insights cannot be tied directly to meditative feats. If this transformation of the person is not attached to a particular state of consciousness, and states of consciousness in turn do not correspond in a simple manner to observable phenomena, the value of physiological studies for the issue of the nature of religious claims is radically undercut.

Can finding a brain-wave change or other physiological basis to mystical experiences then figure at all in determining the truth-value of mystical claims? Obviously something in our physiology *permits* mystical experiences to occur and this can be studied. No one would argue that an experience, scientific or mystical, can involve an insight only if there is no physiological substratum. But the question is whether the insight is determined by our neural or physiological structures. Because of the at best tenuous connection between what physiologists find and insights involved, one would be inclined to think not. Others disagree: if a physical basis were found correlatable with the occurrence of mystical insights, then either the insights are validated (according to advocates of meditation) or discredited (according to critics of religion). Since the occurrence of a mystical insight does happen in a state of consciousness (whether the former is "caused" by the latter or not) and since in principle several complete accounts of any conscious state are possible, one-to-one physiological correlates of the occurrence of any mystical insight may in fact be found. That is, recurring physiological correlates may be found whenever a mystical insight is occurring (although finding an exact correlation may be difficult, as discussed above). Most broadly, the shift of mental functioning in meditation from the left to the right brain-hemisphere (in right-handed people), a shift from linguistic and analytic to nonverbal and synthetic activity, is advanced by some as establishing a physiological basis for mystical claims.[14] Others see any such "natural explanation" (an explanation exclusively in terms of law-abiding "natural" factors such as physiological, psychological, or social ones) as sufficient to undercut any grounds for accepting the mystical claims. William Alston, for instance, would agree "that a successful explanation of certain mystical experiences in terms of purely natural factors would enable us to disallow claims that in these experiences one is directly apprehending God."[15] Thus, for example, the timelessness involved in a mystical experience can be explained as simply due to the lack of content in the mind rather than to the experience of some reality outside of time. The experience can then be fitted into a belief-framework accepted for other reasons, and the mystics' cognitive claims need not be considered.[16]

The position that a natural explanation falsifies the mystics' explanations (or renders them superfluous or implausible) rests on a commitment to a naturalistic metaphysics. No one (except perhaps some mystics) would argue that a mystical explanation—that "ultimate reality" is directly experienced—renders a natural scientific explanation false or implausible. Rather a metaphysical judgment instead of a simple natural explanation (which remains neutral in itself) is involved. Invoking Occam's Razor to disallow reference to factors other than sensory observable ones is question-begging in favor of one metaphysics building up an ontology with material objects as basic. The rigorous application of parsimony leads to solipsism. Ruling out appeal to mystical experiences for cognitive matters is logically of the same order as neutralizing sense-experience for material objects; only differences in metaphysical systems make the situations appear different. And any appeal to parsimony stopping short of solipsism would lead to conflicts as to how reality is actually constructed. In addition, mystical explanations are absolutely simple in their own way (that the nature-mystical whole or the depth-identity rather than a multiplicity of material objects is all that is real). The mystical is radically different in nature from material objects, but not more complex; it offers an overall simpler ontological picture even if it does not explain why one state of affairs is the case rather than another (i.e., the mystical is an alternative to scientific explanations, not part of one).

More integral to the issues of this appendix, again ignoring the issue of whether insights are attached to states of consciousness, both the pro and con positions regarding the role of natural explanations in verifying mystical claims seem difficult to maintain in light of the fact that all conscious states are open to physiological accounts and are *a priori* just as likely to have physiological correlates. EEG examinations of scientists doing research (or for that matter, of Alston while he is thinking through his position) are irrelevant to the truth-value of their claims. Finding one-to-one physiological correlates present whenever scientists are doing experiments or having intuitions that are subsequently shaped into hypotheses does not guarantee a scientific breakthrough or the validity of the theory or experimental results. A "physiology of science" (as opposed to one of scientists) is impossible. Nor would an account of the brain-processes of meditators rule out the possibility of valid insights any more than an account of seeing material objects disproves their existence. If religious believers can accept a causal account of the world,[17] they can also do so here.

William James clearly saw the neutrality of such causal accounts. Scientific theories are "organically conditioned just as much as religious emotions are; and if we only knew the facts intimately enough, we should doubtless see 'the liver' determining the dicta of the sturdy atheist as decisively as it does those of the Methodist under conviction anxious about his soul."[18] In reality there may be no localization of thought—at least no area of the cerebral cortex can be stimulated to cause a person to believe or decide.[19] But even if thinking were correlatable with a biochemical process, still it is most important to realize that we would determine whether a thought is valuable on grounds independent of a physiological basis: we proceed from the established thought to the "proper" physiological basis, not vice versa. It may or may not be that "abnormal" bodily states are necessary to see certain truths about the universe. To use James's example, for all we know, "103 or 104 degrees Fahrenheit might be a much more favorable temperature for truths to germinate and sprout in, than the more ordinary blood-heat of 97 or 98 degrees."[20]

Hence the issue of whether mystical claims are delusions arising from sensory deprivation or are valid insights into the nature of reality will need to be determined on other grounds. The sciences may have established that during some meditation there is no response to external stimuli and that, when not meditating, yogins' sense-experience shows no retention and no habituation from repetition of external stimuli.[21] But to establish that this is seeing reality "as it really is" rather than being just valueless subjectivity will involve weighing different factors and advancing arguments. To illustrate this sort of reasoning, consider possible arguments concerning the left and right brain-hemispheres: (1) Each hemisphere provides a separate sphere of consciousness, and conscious processes grounded in them have equal claims to being insights; therefore, mystical insights are as veridical as ordinary sense-experience even if they cannot be "translated" into ordinary claims. To modify one of Eddington's remarks, if sense-experiences give knowledge of the world, then why should not other experiences produced by a brain developed through natural selection reflect reality too?[22] (2) The right hemisphere's only legitimate function is in conjunction with the left hemisphere in everyday perceptions; recent research speaks less of two separate brains and more of two specialized parts functioning together in a highly integrated manner.[23] Thus a mystical experience is merely a malfunction, a useless spinning of gears incapable of a true insight into the workings of the world. This latter position is not the direct negation of the former, but these positions reveal that physiological data are themselves neutral and open to interpretation.

That there have been physiological studies of meditation for more than forty years and the philosophical issues connected with it have not been settled should therefore not appear surprising. The most that the sciences can establish is that there are different states of consciousness, not what value, if any, there is in the insights that may or may not be tied to them. Establishing a difference in the electrical activity of the brain or a lack of it simply proves nothing. There are no "crucial experiments" in this area. To speak of scientific *evidence* in such circumstances is inappropriate if experiments cannot figure in deciding the issues.

The general conclusion from this discussion must be that there is no physiological study of religious and mystical insights, and consequently that in the most important sense there is no empirical, scientific study of religious and mystical dimensions to experiences.[24] The traditional "genetic fallacy" argument (that accounts of the origin or cause of belief are irrelevant to the truth or falsity of that belief) is not applicable here since the correlated bodily or conscious states (if such correlates are found) are not necessarily the cause. So too the various foci (body, state of consciousness, and insight-experience) should be distinguished in order to see the full complexity of the situation.

If the sciences are neutral on the matter of the belief-component, why then is there so much interest in "physiological studies of the mind"? One factor is a general acceptance of intersubjective testability (scientific objectivity) as a cognitive standard: what can be publically reproduced is considered more "real" than the subjective (experiential) phenomena. Behaviorialism, the attempt to understand behavior without appeal to "inner episodes," is one product of this tendency.[25] A second factor is the interest in the "secular" effects of meditation: Benson, as his other books show, and others are concerned only with the possible beneficial or harmful effects meditation may have upon the body and personality (e.g., lowering blood-pressure or lessening stress). "Subjective" matters related to enlightenment-claims are inconsequential to them: any meditator, whether undergoing a permanent change in his or her belief-framework or not, is of equal importance to them; if anything, those who have achieved paranormal powers *(siddhis)* or especially great levels of concentration—feats in themselves not necessarily related to the insight-experiences—are most important.[26] For their interests, physiological correlates are of value, but meditation is thereby reduced to mere health techniques. The claims that studies of these effects provide a "Western understanding of the spiritual" or a "physics of religion" or even a "scientific religion" rest upon this reduction. Concern for stress or the nature of perception is legitimate and valuable, but it should not be equated with the religious concerns (ending all suffering, gaining absolute knowledge of reality, or whatever).

Scientific versus Socioscientific Studies of Religion

The nature of this natural scientific approach to mystical experiences can be brought out more fully by a comparison to socioscientific approaches to religiosity in general. For example, the difficulty of correlating states of the brain with religious insights is roughly paralleled by the difficulty of correlating a "state" of society with an *ethos* and with a religious "reaction": the decline of the Chou dynasty in China produced two significantly diverse religious traditions (that of Confucius and that of Lao tzu); conversely, two different sets of social conditions can produce the same religious phenomenon, if we are to believe the theories that "Buddhism was the product of poverty" and that "Buddhism was the product of prosperity."

More important, however, are the differences between the two types of approaches. In constructing theories, in principle social scientists are concerned only with religion as a factor of social life and its relation to other social realities; as a science, socioscience "deals with relations, not with origins or essences."[27] But in practice one major socioscientific approach drawing its inspiration from such thinkers as Émile Durkheim (who thought that the social group was the only reality behind the concept "God") proceeds not by discovering simple correlations between religious and other factors but by giving greater weight to other cultural, social, or psychological factors in order to develop an explanation of the "surface" (religious) phenomena.[28] Giving priority to the nonreligious components—making them more "real"—is comparable, not to *scientific* explanations, but to giving the brain-correlates greater weight in *metaphysical* positions on the mind-body problem.[29] This epiphenomenalist position with regard to religion has been a "fundamentalism" of academia in this country.[30] To give an instance of this, Edmund Leach says that Kachin spirits are "in the last analysis *nothing more* than ways of describing the formal relationships that exist between *real* persons and *real* groups in ordinary Kachin society."[31] So also myths from all cultures, which ostensibly are about the creation of the world or the interaction of gods and humans, are seen by others as really no more than symbols of whatever psychosocial phenomenon is taken to be fundmental.[32] What these social scientists deem real dictates what is an "objective" description or explanation.

By valuing social or psychological forces as more fundamental, religion can be "reduced" or "explained away" in the sense that the only reality behind religious concepts is ultimately a social or psychological factor, not that to which the concepts appear to refer, or that religion is directly and absolutely dependent upon psychosocial factors if it is not treated as totally ineffective in our lives. The cognitive content of a believer's faith is thereby discredited; believers are radically mistaken as to the actual forces of reality. Scientific explanations are possible without reducing any potential insights, as with physiology. Reductionism involves a philosophical position that goes beyond finding scientific correlates of a phenomenon to ruling out the possibility that the phenomenon may involve a unique insight into reality. Thus the proper understanding of religious phenomena's true cognitive content is provided by other disciplines. In the area of science this does not compare with the reduction of one theory to another (i.e., where the phenomena explained by one theory are now explained by a theory referring to more fundamental entities). A scientific reduction involves correlating, say, biological processes with chemical or physical condition: there is no implication that the biological explanations are not necessary and insightful for its level of organization. Socioscientific distinctions between factors that are real and those which are not set this approach apart from natural science.[33] Religious explanations of phenomena are discredited in reductionist socioscientific explanations in a way that, say, biological explanations of phenomena are not discredited by quantum physics. Instead, the socioscientific contention that religious claims are fundamentally misguided is comparable to Advaita Vedānta's "explaining away" the world. To the extent that socioscientific approaches to religiosity reductively equate psychosocial phenomena with the referent of the religious meaning, the understanding provided by religious believers will necessarily conflict with that provided by these approaches.[34] In other words, if the only reality in a cultural phenomenon is biological or psychosocial, then the participant's explanation will be wrong in principle.

If the referents of religious concepts were reducible to physiological or socioscientific realities in a sense analogous to the way scientific ones are, religious phenomena nonetheless would still have to be treated as unique in the way biological processes are.[35] Melford Spiro thinks no religious believer should object to the reduction of religion to its "chemical constituents."[36] If the scientific idea of explanatory reduction is meant, this is true. It is equivalent, modifying his example, to a scientific analysis of the paints and canvas of a painting. Such an analysis does not destroy the "essential quality" of art. But it is objectionable to refer to this as a *scientific* analysis of *art* since it misses entirely the aesthetic dimension (the "essential quality" of art). Something connected to art—the materials utilized—is being studied, but not the art itself.

The consequence from such claims as Leach's that "ritual action and belief are alike to be understood as forms of symbolic statements about the social order"[37] is to shift the focus of attention away from a religious dimension to exclusively socioscientific real factors by equating religious statements with symbolic statements about these phenomena. This is similar to maintaining that we are studying religious insights by studying the physiological effects of meditation: religious statements would become merely symbolic statements of changes in the electrical activity of the brain or any other physiological process accessible to scientific investigation; ultimately, the true referent of each type of statement (religious and physiological) would be the same.[38] Problems of even correlating insights with brain-states were raised earlier.

Unlike physiological accounts of meditation, socioscientific explanations in this one approach are always taken as inflicting irreparable damage to religious claims. Assuming that religious experiences or beliefs can be correlated with psychological or social conditions is itself problematic, as mentioned earlier. In physiology, whether an experience can be correlated with physiological data is an open question. This socioscientific approach, however, is more metaphysical, tending to fit any data within its various particular systems. Falsification is much more difficult, if not impossible.

But once again the fact remains that all experiences and beliefs are in principle open to different such accounts (political, economic, and so on). To take an example from another enterprise making claims about the world based upon experiences, the natural sciences: Newton's conception of an absolutely mechanical world may have been the projection of a father who governed in a rigidly predictable manner and with absolute authority, or may have been compensation for a lack of such a father. Still the Newtonian framework was replaced with relativity and quantum mechanics by recalcitrant anomalies and the creative thinking of physicists, not by a psychological analysis of Newton's early life. Religious claims about the world are treated differently only because of the belief that religious ideas are fundamentally erroneous—something not derivable from the socio-scientific studies themselves.[39] Sociologists of science, while exploring the influence of society upon the development of science, never treat scientific claims about the world as symbolic statements whose actual referents are real social forces. Statements from the natural sciences are never given socioscientific reductions any more than they are dismissed as a dysfunction of the brain because they are the product of the extreme specialization of the left brain-hemisphere (or whatever). For either case, arguments must be advanced independent of the empirical findings to support what is a *prima facie* absurd thesis concerning alleged scientific and religious insights. If scientific claims are exempted from reduction because they, say, are empirically testable, then at least a complete and final shift in the standard of validation away from psychosocial causes and in the direction of argument is being admitted.

It might be argued that mystical experiences are necessarily less veridical than ordinary ones because mystics are not all socially or psychologically well-adjusted, although defining psychosocial "health" is much more difficult than defining physical health and therefore is more apt to be question-begging. From a psychological point of view, certainly anyone who would want to overcome a sense of self could only be classified as pathological.[40] And major differences in values might also put mystics out of step with society at large. Crises and imbalances may have impelled them along the way of life they choose.[41] But psychosocial causes and functions are themselves as irrelevant to the possible truth of the claims involved as physiological ones. Problems here can be seen with examples again from the natural sciences: in some instances, the interest of scientists in the repeatable and the predictable may reveal a neurotic fear of the unique, the unknown, the erratic, and the unexpected, as Abraham Maslow observed.[42] No one would consider this sufficient grounds for dismissing all of science. Or more concretely: Newton is known to have been vain, ambitious, and competitive. In fact, it could have been these very drives that were instrumental in his becoming one of the greatest scientists of all time. Here equating "true" with "arising from a psychologically/socially healthy state" would be unwarranted, and, without an additional argument, the same is true elsewhere.

Reasons and Causes

Those anthropologists, sociologists, and psychologists who have built-in ontological assumptions go beyond producing merely empirical studies to usurping the role of philosophy: they recognize the causes and functions constituting natural explanations as the sole grounds for evaluating the possible truth of religious claims. Replacing *reasons* with *causes* would destroy all positions concerning religious faith as well as science.[43] A primary difficulty with this position is that even recognizing a certain process as the cause would need to be argued for. The causes and the personal or social needs that religious beliefs satisfy may explain why religions *persist* or *perish*, but the above problems preclude their relevance for the issue of whether religious claims are true or not. If religious claims were unprovable, still the presence of a natural explanation would not figure in accepting or rejecting them, since all beliefs are so analyzable. We may reject a claim if we have other reasons to doubt it (e.g., if it conflicts with a deeply held belief) and thus may want to explain it away by appeal to natural explanations; that is, we would be satisfied that nothing other than what we are committed to as being real is actually involved. But "explaining away" beliefs involves this appeal to argument; the natural explanations themselves would remain neutral.

Psychosocial factors might be thought to enter into evaluating the cognitive content of a religious faith, as with finding psychological conditions explaining why Einstein so strongly favored simplicity in scientific theories. Yet this is only indirect; reasons would still need to be found why simplicity should be considered among factors at all in evaluating scientific theories, and there remain the difficulties in determining which is actually the simplest theory once simplicity is admitted as a value. For religious claims, psychologically favored factors need to be treated similarly: reasons are the final court of appeal in determining the significance of a religious experience.

At best, through a relentless inquiry into the foundations of belief—that is, first giving a reason for a belief, then giving a reason for that reason, and so forth—we might reach a point where agreed-upon background-beliefs are exhausted and consequently where a decision or judgment must be made for which no reasons acceptable to all can be given. Ludwig Wittgenstein felt that ultimately all our believing is thus groundless, neither reasonable nor unreasonable.[44] If this is so, values conditioned by psychosocial causes would need to be invoked. What is advanced as "self-evident" at this level may depend upon such causes. If this situation does occur, all our beliefs, not just scientific, religious, and anti-religious ones, would be affected. At this fundamental level, causes for accepting a claim and the reasons advanced for believing it to be true would not be distinct. Unlike in the physiological situation, psychosocial conditions may enter into the insights themselves.[45]

Conclusion

There are no purely religious phenomena: every phenomenon has multiple dimensions. Religious groups are groups shaped by real social forces (as are scientists); religious meditation has physiological elements shaped by the nature of the body. Myths and religious cosmologies do rely upon natural, social, and psychological phenomena as analogies for their interpretations of the sacred; they are constructed out of the material at hand as much as ritual implements are.

If the metaphysically loaded position of the socioscientists discussed here concerning the real factors involved in cultural phenomena is rejected, socioscientific studies of religion can be legitimately modeled upon physiology. That is, necessary and sufficient conditions for the emergence of a particular unobservable religious experience or belief can be specified (if they are found), but no metaphysical causal priority between the correlated conditions and the "subjective" element is assumed. In physiology, theories and causation are restricted to the order within and between each type of physical correlate (muscles, nerves, and so on); no metaphysical causation between consciousness and matter needs to be invoked. So too social scientists can study the correlation of religious phenomena with other human phenomena without assuming a causal connection. Psychosocial factors would be the necessary and sufficient conditions for the emergence of the religious insights (and vice versa). Religious ideas would be competing in the marketplace of ideas with other cultural ideas, and observable religious behavior would be one cultural factor interacting with the other elements of a culture. Different theorists may disagree on how much weight to give each factor, but as long as no factor is totally discredited in advance, the theorists would be arguing over "scientific," not "metaphysical," matters. Problems of a metaphysical sort enter the situation only when the physical, social, or psychological correlates of religiosity are identified with its "essential quality." Because there is as yet no objective means to study the experiential or transformative dimension of mysticism, what is observable—what appears to the public—is not the only datum of interest. The religious dimension can be studied only by focusing upon the believers' descriptions of their faith. Their claims need not be accepted uncritically, but a religious dimension must not be erased beforehand. Believers may be wrong or unjustified in their faith, but they are no more rationalizing in principle than are scientists. Claims about the nature of the world and the sacred should be treated as precisely that: without an appeal to metaphysics we cannot reduce a believer's "ultimate concern" to an observer's concerns, whether those concerns are social, psychological, or physiological.

Notes

Introduction

1. For a discussion of science and philosophical Taoism, see Richard H. Jones, "Mysticism and Science: Against Needham on Taoism," *Journal of Chinese Philosophy* 8 (1981): 245–266.

2. Having mystical experiences would not be sufficient for this study; a careful study of the texts would still be required to understand how the Buddha and Śaṁkara construed reality, since mystics do not all see reality in the same way.

3. Werner Heisenberg, *Physics and Philosophy* (New York: Harper and Row, 1958), p. 187.

Chapter 1. Science

1. Albert Einstein, "Autobiographical Notes," in P. A. Schilpp, ed., *Albert Einstein: Philosopher-Scientist* (LaSalle, Ill.: Open Court, 1949), p. 9. The lucidity and certainty of geometry also deeply impressed him.

2. See Jerome R. Ravetz, *Scientific Knowledge and its Social Problems* (New York: Oxford University Press, 1971).

3. Stephen Toulmin, *The New York Review of Books* 24, no. 10 (June 9, 1977): 4.

4. It makes sense to speak of seeing "sense-data" when persons blind from birth have their sight restored but have no conceptual schemes to classify data—before they learn to see, theirs is a world of color-patches only—or in science when there is a dispute over the interpretation of visual data. The scientifically significant problems unrelated to the nature of observation enter the picture only with the introduction of object-language, as occurs in all normal observation.

5. Willard Van Orman Quine, *The Ways of Paradox and Other Essays* (New York: Random House, 1966), p. 211.

6. The phrase comes from Stephen Toulmin, *Foresight and Understanding* (New York: Harper and Row, 1961).

7. Hermann Bondi, *Assumption and Myth in Physical Theory* (Cambridge: Cambridge University Press, 1967), pp. 79–80.

8. See Carl Hempel, *Aspects of Scientific Explanation* (New York: The Free Press, 1965) for a defense of this. He also accepts probabilistic explanations.

9. Toulmin, *Foresight and Understanding;* N. R. Hanson, "On the Symmetry between Explanation and Prediction," *Philosophical Review* 68 (1959): 349–58.

10. Conclusion to *Mathematical Principles of Natural Philosophy.*

11. Pierre Duhem, *The Aim and Structure of Physical Theory* (New York: Atheneum, 1974), pp. 55f.

12. Ernest Nagel, *The Structure of Science* (New York: Harcourt, Brace and World, 1961), p. 107.

13. Werner Heisenberg, *Physics and Philosophy*, p. 168.
14. Carl Hempel, *Aspects of Scientific Explanation*, pp. 430–33.
15. Willard Van Orman Quine, *From a Logical Point of View* (New York: Harper and Row, 1953), pp. 42–43.
16. Hempel, *Aspects of Scientific Explanation*, p. 95.
17. Cited in Robert E. Ornstein, *The Mind Field* (New York: Pocket Books, 1978), p. 85.
18. Susanne K. Langer, *Philosophy in a New Key* (New York: The New American Library, 1952), p. 227.
19. Karl Popper, *The Logic of Scientific Discovery* (New York: Harper and Row, 1959), p. 280.
20. N. R. Hanson, *Observation and Explanation* (New York: Harper and Row, 1971), p. 12.
21. N. R. Hanson, *Patterns of Discovery* (Cambridge: Cambridge University Press, 1958), p. 54.
22. Duhem, *The Aim and Structure of Physical Theory*, pp. 183–88.
23. Ibid., pp. 190–95.
24. Contra Thomas Kuhn, *The Structure of Scientific Revolutions* (Chicago: University of Chicago Press, 1969), 2d ed.; and Paul Feyerabend, *Against Method* (London: NLB, 1975).
25. Imre Lakatos, "Falsification and the Methodology of Scientific Research Programmes," in Imre Lakatos and Alan Musgrave, eds., *Criticism and the Growth of Knowledge* (Cambridge: Cambridge University Press, 1970), p. 92.
26. Nagel, *The Structure of Science*, pp. 116–17.
27. A philosophical formalization of scientific theories (a logical reconstruction of premises and conclusions) at best reveals logical consistency. It may reveal problems with a theory but cannot add anything more positive toward the justification of a theory since false theories can be consistent too.
28. Michelson of the Michelson-Morley experiments (which supposedly once and for all discredited the ether theory) went on believing in ether until his death—some forty years after the first publication of results.
29. Popper, *The Logic of Scientific Discovery*, p. 50.
30. Stephen Toulmin, "Philosophy of Science," *Encyclopedia Britannica* (15th ed.) 16: 383.
31. Popper, *The Logic of Scientific Discovery*, p. 105.
32. Imre Lakatos and Alan Musgrave, eds., *Criticism and the Growth of Knowledge* (Cambridge: Cambridge University Press, 1970), pp. 119, 179.
33. Cited in Kuhn, *The Structure of Scientific Revolutions*, p. 151.
34. Mary Hesse, *The Structure of Scientific Inference* (Berkeley/Los Angeles: University of California Press, 1974), pp. 35–36.
35. See Feyerabend, *Against Method*, and *Science in a Free Society* (London: NLB, 1978).
36. Steven Weinberg, *The First Three Minutes of the Universe* (New York: Basic Books, 1977), p. 119.
37. Feyerabend, *Against Method*, pp. 284–85, 303.
38. Popper, *The Logic of Scientific Discovery*, p. 37.
39. Kuhn, *The Structure of Scientific Revolutions*, pp. 205–7.
40. Hesse, *The Structure of Scientific Inference*, p. 117.
41. Henri Poincaré, *Science and Hypothesis* (New York: Dover, 1952), p. xxiii.
42. These should not be confused with ideal entities such as frictionless planes or treating our solar system as a dimensionless point-mass. Such entities are abstractions that are merely simpler than the literal complex truth. They are useful in analyzing real situations without the implication that they are part of the actual workings of the world.
43. Arthur Eddington, *The Nature of the Physical World* (Ann Arbor: University of Michigan Press, 1974), p. 244.
44. Mario Bunge, "The Weight of Simplicity in the Construction and Assaying of Scientific Theories," in M. Foster and M. Martin, eds., *Probability, Confirmation and Simplicity* (New York: Odyssey Press, 1966), p. 308. Nancy Cartwright in *How the Laws of Physics Lie* (New York:

Oxford University Press, 1983) would go farther in claiming that all constants and regularities described in theories (while caused by theoretical entities) are only the result of abstraction.

45. Nagel, *The Structure of Science,* p. 366.

46. See Gerald Holton et al., "Do Life Processes Transcend Physics and Chemistry?" *Zygon* 3 (1968): 442–72.

47. Gerald Feinberg, *What is the World Made of?* (Garden City: Doubleday, 1978), p. 262.

48. Alan W. Watts, *Psychotherapy East and West* (New York: Ballantine Books, 1961), p. 17.

49. Nelson Goodman, "The Way the World Is," *Review of Metaphysics* 14 (1960): 48–56.

50. Gilbert Ryle, *Dilemmas* (Cambridge: Cambridge University Press, 1954), pp. 68–81.

Chapter 2. Mysticism

1. William A. Lessa and Evon Z. Vogt, eds., *Reader in Comparative Religion* 3d ed. (New York: Harper and Row, 1972), p. 1.

2. See Clifford Geertz, *Islam Observed* (Chicago: University of Chicago Press, 1971), pp. 97f.; Arthur C. Danto, *Mysticism and Morality* (New York: Harper and Row, 1972), pp. viif.

3. See Ninian Smart, "Interpretation and Mystical Experience," *Religious Studies* 1 (1965): 75–76 for a similar distinction. (The phrase "nature-mystical" is adapted from the English nature-poets but covers certain introspective experiences such as a sense of union.)

4. Mystics often do not consider enlightenment an "experience"; it is an *insight,* a shift in knowledge that cannot be equated with any experience. But whatever the cause of the insight, it occurs in a relatively short time-span; thus it can legitimately be referred to as an experience.

5. Charles T. Tart, *States of Consciousness* (New York: Dutton, 1975), p. 97.

6. Peter L. Berger, "Identity as a Problem in the Sociology of Knowledge," in *The Sociology of Knowledge: A Reader,* ed. James E. Curtis and John W. Petras (New York: Praeger, 1970), p. 376.

7. Tart, *States of Consciousness,* pp. 84–85.

8. Peter L. Berger, "Identity as a Problem in the Sociology of Knowledge," p. 375, discussing the socialization of everyone.

9. Robert E. Ornstein and Claudio Naranjo, *On the Psychology of Meditation* (New York: Viking, 1971), p. 196. In such meditation, all structuring elements are removed.

10. Smart, "Interpretation and Mystical Experience," p. 87.

11. There is the overwhelming sense of reaching "ultimate reality"—that there is no additional or more fundamental reality to be known. But mystics cannot circumvent the basic epistemological problem of leaping from experiences to ontological claims about the world independent of the experience, as differences in the understanding of the status of the mystical indicates.

12. For the implications of this for the moral status of this way of life see Richard H. Jones, "Theravāda Buddhism and Morality," *Journal of the American Academy of Religion* 47 (1979): 371–87.

13. Edward Conze, "Buddhist Philosophy and its European Parallels," *Philosophy East and West* 13 (1963): 11. On *"nibbāna"* as a religious concept, see Guy Welbon, *The Buddhist Nirvana and its Western Interpreters* (Chicago: University of Chicago Press, 1968), pp. 302, 304.

14. Henry Clarke Warren, *Buddhism in Translation* (New York: Atheneum, 1972), p. 173.

15. Ibid., p. 6.

16. Ibid., p. 316.

17. The Buddha was not opposed to suicide if it was accomplished with a complete lack of craving for existence or for freedom from existence (M. 3.263, S 1.120, S 3.119, S 4.55).

18. *"Kamma," "nibbāna,"* and "Brahman" will remain untranslated. Although the concepts are explainable in English, there are no simple English equivalents for them. *"Kamma"* literally means "deed," but it has such a specialized meaning that substituting the term "deed" for it would be misleading.

19. Winston L. King, "The Structure and Dynamics of the Attainment of Cessation in Theravāda Meditation," *Journal of the American Academy of Religion* 45 (1977): 659, 670; on the Buddha's enlightenment and the trances, pp. 661, 664.

20. There are also feelings in this state, but no craving or attachments to them arise (S II.82).

21. William Peiris, *The Western Contribution to Buddhism* (Delhi: Motilal Banarsidas, 1973), pp. 98–99. Peiris also cites Conze's endorsement of this: "The more I am concerned with these things, the more convinced I become that George Grimm's interpretation of the Buddhist theory of *ātman* comes nearest to the original teaching of the Buddha."

22. Edward Conze, *Thirty Years of Buddhist Studies* (Columbia: University of South Carolina Press, 1968), p. 11.

23. Because Śaṁkara's work is mostly in the form of commentaries, the religious problem of suffering and rebirth is not always explicit.

24. Tart, *States of Consciousness*, pp. 26–32. Some physiologists (such as Wilder Penfield) see the brain as a conduit for consciousness.

25. Modified from Śaṁkara's example (BSB III.2.11) of a clear crystal, a red color as limiting adjunct, and taking the adjunct as real.

26. Translated by George Thibaut, *The Vedānta Sūtras of Bādarāyaṇa* (New York: Dover, 1962), p. 4. The following points are from BSB, Introduction.

27. See Daniel H. H. Ingalls, "Śaṁkara on the Question: Whose is Avidyā?" *Philosophy East and West* 3 (1953): 69–72, for a discussion.

28. Contra Franklin Edgerton "The *Upaniṣads:* What do they seek, and why?" *Journal of the American Oriental Society* 49 (1929): 97–121.

29. For early Upaniṣadic thought, the knower does not become nonexistent in dreamless sleep but merely inactive (because it is without an object) (cf. BU IV.3.23).

30. For more on this interpretation, see Richard H. Jones, "*Vidyā* and *Avidyā* in the *Īśa Upaniṣad*," *Philosophy East and West* 31 (1981): 79–87.

31. BU II.4.5 does say the self is to be seen and reflected upon—by this, all the world is known. This is similar to seeing the reality beneath name and form in the *Chāndogya.* But the *Bṛhadāraṇyaka* passage goes on to say that we cannot know the knower of knowing, etc.

32. Śaṁkara, *Crest-Jewel of Discrimination,* trans. Swami Prabhavananda and Christopher Isherwood (New York: New American Library, 1970), pp. 75, 85.

33. Action is based on distinguishing actor and action, and action and result, while knowledge is a correction involving no action.

Chapter 3 Ways of Life and Knowledge-Claims

1. See Stephen E. Toulmin, "Contemporary Scientific Mythology," in Stephen E. Toulmin et al., *Metaphysical Beliefs* (New York: Schocken, 1957), pp. 37–56; and Anthony Quinton, "Ethics and the Theory of Evolution," in *The Sociobiology Debate,* ed. Arthur L. Caplan (New York: Harper and Row, 1978), pp. 117–41.

2. Weinberg, *The First Three Minutes of the Universe,* p. 154.

3. Eddington, *The Nature of the Physical World,* pp. 85–86.

4. Robert Jastrow, *God and the Astronomers* (New York: W. W. Norton, 1978), p. 16.

5. Ibid., p. 113.

6. Lincoln Barnett, *The Universe and Dr. Einstein* (New York: Time-Life, 1948), p. 100. Einstein's "cosmological awe" involves the "deep conviction of the rationality of the universe" and a "yearning to understand it," not traditional "religious" or "mystical" desires.

7. See Robert K. Merton, *The Sociology of Science,* ed. Norman W. Storer (Chicago: University of Chicago Press, 1973). But this ideal, of course, is not always fulfilled; see William Broad

and Nicholas Wade, *Betrayers of the Truth: Fraud and Deceit in the Halls of Science* (New York: Simon and Schuster, 1982).

8. Imre Lakatos, *Proofs and Refutations,* ed. John Worrall and Elie Zahar (Cambridge: Cambridge University Press, 1976), pp. 142–43.

9. Richard Rudner would go as far as to argue that scientists *qua* scientists make *moral* judgments in deciding the *acceptability* of a theory; see his "The Scientist *Qua* Scientist Makes Value Judgments," *Philosoophy of Science* 20 (1953): 1–6.

10. There is a tension between mindfulness and the attention to distinctions needed for scientific research. Once far enough along the path to enlightenment or once enlightened, mystics may not value science or be concerned about increasing the details of knowledge within "the dream." In chapter 8 the general rejection of science by mystics will be discussed.

11. Peter Berger, *The Sacred Canopy* (Garden City: Doubleday, 1967), p. 22.

12. Clyde Kluckhohn, "Myths and Rituals: A General Theory" reprinted in *Reader in Comparative Religion: An Anthropological Approach,* ed. Lessa and Vogt, pp. 93–105.

13. Mary Douglas, "Pollution," reprinted in ibid. pp. 196–202.

14. For example, Erich Fromm, *Psychoanalysis and Religion* (New York: Bantam, 1967), p. 22; Huston Smith "The Death and Rebirth of Metaphysics" in *Process and Divinity,* ed. W. L. Reese and E. Freeman (LaSalle: Open Court, 1964), p. 41.

15. Robin Horton, "African Traditional Thought and Western Science" reprinted in *Rationality,* ed. Bryan R. Wilson (New York: Harper and Row, 1970), pp. 131–71.

16. Langer, *Philosophy in a New Key,* p. 4.

17. Peter L. Berger and Thomas Luckmann, *The Social Construction of Reality* (Garden City: Doubleday, 1967), p. 80.

18. Huston Smith, "Tao Now: An Ecological Testament," in *Earth Might Be Fair,* ed. Ian G. Barbour (Englewood Cliffs: Prentice-Hall, 1972).

19. For an illuminating discussion, see R. Hooykaas, *Religion and the Rise of Modern Science* (Grand Rapids: William B. Eerdmans, 1972).

20. See Bronislaw Malinowsky, *Magic, Science and Religion* (Garden City: Doubleday, 1954) on science, i.e., empirical knowledge.

21. Individual mystics may have had occult interests of importance to the development of modern science incorporated in their mystical ways of life, but the specifically *mystical* interests and practices are distinct. Mysticism is not the "cause" of science in any simplistic manner. It cannot be the sufficient condition for science's development since there has been mysticism in the West during periods when scientific development was dormant. So too, if it were sufficient, a modern science should have developed in India, a country whose scientific innovations in classical times did not match its mystical ones. Nor is mysticism a necessary condition: some concern for experience is necessary for the development of science, but mystical experiences may be antithetical to scientific ones (as will be discussed in chapter 6).

Chapter 4. Reality

1. Term from Stephen C. Pepper, *World Hypotheses* (Berkeley/Los Angeles: University of California Press, 1942).

2. Pepper doubts the adequacy of mysticism as a world-view because so much is denied as unreal (see *World Hypotheses,* pp. 127–35).

3. To illustrate a difference in metaphysics, it can be noted that neither the Advaitins nor the Theravādins conceive of a mind/body problem as is done in the West. Material versus mental properties is not an important contrast here. The Buddha went as far as to say that if we were to identify ourselves with either the mind *(citta)* or the body *(kāya),* it should be with the latter since it lasts for years while the mind changes constantly (S II. 94–97).

4. A point made by J. O. Wisdom, "Scientific Theory: Empirical Content, Embedded

Ontology, and Weltanschauung," *Philosophy and Phenomenological Research* 33 (1972): 70–72; and J. W. N. Watkins, "Confirmable and Influential Metaphysics," *Mind* 67 (1958): 355–57.

5. J. F. Staal, *Advaita and Neo-Platonism:* (Madras: University of Madras, 1961), p. 127.

6. Melford E. Spiro, *Buddhism and Society* (New York: Harper and Row, 1970), p. 141; Alex Wayman, "Buddhist Dependent Origination," *History of Religions* 10 (1971): 197–98.

7. Term from Peter F. Strawson, *Individuals* (Garden City: Doubleday, 1963), p. xiii.

8. Lama Anagarika Govinda, *The Psychological Attitude of Early Buddhist Philosophy* (New York: Samuel Weiser, 1974), p. 39.

9. Thus the enlightened are said to reach voidness *(sunnatā)*, signlessness *(animittatā)*, and the desireless *(appaṇihitā)* (S IV. 360).

10. Nyanaponika Thera, *The Heart of Buddhist Meditation* (New York: Samuel Weiser, 1973), p. 171.

11. In the Abhidhamma, Nyanaponika notes, it is believed that even the earliest stage of bare sense-experience carries a subtle "flavor" from earlier impressions (as with habit-energy)—even in "bare attention" (ibid., pp. 36, 112).

12. K. N. Jayatilleke, "Buddhism and the Scientific Revolution," in *Pathways of Buddhist Thought*, ed. M. O'C. Walshe (London: George Allen and Unwin, 1971), p. 99. Pain may be essential for survival, but the Buddha ultimately does not value that.

13. Nyanaponika, *Heart of Buddhist Meditation,* p. 171.

14. Arthur Deikman, "Deautomatization and the Mystic Experience," reprinted in *The Nature of Human Consciousness,* ed. Robert E. Ornstein (San Francisco: W. H. Freeman, 1973), pp. 216–33.

15. In the mystical system as a whole, even the enlightened Buddha talks of past deeds and their effect upon future events and rebirths. Only in mindfulness exercises is there a lack of these concerns.

16. See Sidney Morgenbesser, "The Realist-Instrumentalist Controversy," in *Philosophy, Science and Method,* ed. Sidney Morgenbesser et al. (New York: St. Martin's Press, 1969) on "scientific realism."

17. Ernest Gellner, *Legitimation of Belief* (Cambridge: Cambridge University Press, 1974), p. 63.

18. Poincaré, *Science and Hypothesis,* p. 140.

19. Ian G. Barbour, *Issues in Science and Religion* (New York: Harper and Row, 1966), p. 173.

20. The Theravāda and other mystical traditions also discuss explanatory mechanisms (e.g., *kamma* and dependent origination). They do not stick to the "surface" alone. This will be discussed below.

21. Duhem, *The Aim and Structure of Physical Theory,* p. 7.

22. See Robert E. Ornstein, *On the Experience of Time* (Baltimore: Penguin Books, 1969) for a possible explanation of this.

23. See Olivier Costa de Beauregard, "Time in Relativity Theory: Arguments for a Philosophy of Being," and Milač Čapek, "Time in Relativity Theory: Arguments for a Philosophy of Becoming," in *The Voices of Time,* ed. J. T. Fraser (New York: George Braziller, 1966), pp. 417–33 and 434–54.

24. Cited in ibid., p. 580.

25. For example, Kurt Gödel, "A Remark about the Relationship between Relativity Theory and Idealistic Philosophy," in *Albert Einstein: Philosopher-Scientist,* ed. Paul A. Schilpp (La Salle, Ill.: Open Court, 1949), pp. 557–62. Nothing in this position justifies the idea of changing the past in any way.

26. Mircea Eliade, "Indian Symbolisms of Time and Eternity," in his *Images and Symbols* (New York: Sheed and Ward, 1969), pp. 70–71.

27. Ibid., p. 91.

28. We need not hypothesize on account of precognition and retrocognition (assuming they are genuine) that the past and future are real in the sense of being present and affecting the course of events today. This is so because the theory of *kamma* offers a simpler alternative:

through the paranormal powers, we can retrace the kammic seeds present in a "person" today back to their origin and project them forward to their probable fruition. Nothing more closely linked to the notion of time-travel is listed among the *siddhī.*

29. According to Advaita, as with the Theravādins and most other Indian systems, the cycle of rebirths is beginningless (BSB II.136).

30. The *ontological* matter of the relation of the temporal and atemporal dimensions of reality that come together in the experience of the mystical is more difficult. Thus Advaita would deny the reality of the interval of time while science would assume it.

31. Albert Einstein, *The Meaning of Relativity* (Princeton: Princeton University Press, 1922), p. 55.

32. *"Ākāśa"* is sometimes translated as "ether" to indicate that involved here is not merely the lack of something (as in the phrase "empty space") but a reality present when other things are absent. Some Indian traditions differentiate the eternal directional space *(dik)* from the noneternal medium of sound *(ākāśa)* (BSB II.3.7). But in scientific contexts the term "ether" has gained the restricted sense of the construct posited for one particular problem—the need for a medium through which electromagnetic radiation could pass, analogous to the need for a medium for other wave-phenomena. Since *"ākāśa"* has nothing in common with this scientific posit, "space" is a preferable, more neutral translation.

33. *"Ākāsa"* comes up in one other context in Theravāda texts—that of the "realm of infinite space," one of the higher states of concentration. This is not an endorsement of the infinity of space (a subject unanswered by the Buddha), but involves concentrative meditative exercises only, and not the techniques related to mindfulness and enlightenment (*vipassanā*-cultivation).

34. See Y. Karunadasa, *Buddhist Analysis of Matter* (Colombo: Department of Cultural Affairs, 1967), pp. 93–94.

35. See N. R. Hanson, *Patterns of Discovery,* pp. 50–69.

36. Edwin A. Burtt, *The Metaphysical Foundations of Modern Physical Science* (Garden City: Doubleday, 1954), p. 228.

37. According to the most popular interpretation of quantum physics—the Copenhagen Interpretation—atomic systems *inherently* contain acausal events. If this theory is correct, strict causation would be limited to higher levels of reality.

38. Rudolf Carnap, *An Introduction to the Philosophy of Science,* ed. Martin Gardner (New York: Basic Books, 1966), p. 10.

39. K. N. Jayatilleke, *Early Buddhist Theory of Knowledge* (London: George Allen and Unwin, 1963), p. 443.

40. See Karl Potter, *Presuppositions of India's Philosophies* (Englewood Cliffs: Prentice-Hall, 1963), chap. 1. In the Theravāda conception of enlightenment, a final "leap" after the "path" is necessary, since no action or accumulation of merit can produce enlightenment.

41. The Ājīvikas, with their denial of karmic efficiency, were fatalists.

42. Enlightenment cannot be produced by our volitional acts—hence it is called "unconstructed" *(asaṅkhāra)*—nor do the acts of the enlightened product kammic fruits. *Kamma* involves only acts generated by hatred, greed, and delusion—not all acts. This limitation of range does not affect the issue of whether *kamma* is causal or merely lawful in general.

43. Some philosophers would in fact argue that such lawfulness is *necessary* for free will—without the predictability of outcome, it is hard to understand the notion of choosing.

44. Eliot Deutsch, *Advaita Vedānta: A Philosophical Reconstruction* (Honolulu: The University Press of Hawaii, 1969), p. 69.

45. Ninian Smart, *Doctrine and Argument in Indian Philosophy* (London: George Allen and Unwin, 1964), pp. 51–52.

46. Dependent origination does not enumerate *all* the conditions necessary for each next step. Thus, for a birth to occur, sperm, an ovum, and the kammic residue from the past *(viññaṇa,* the "consciousness" connecting rebirths) must all be present (M I.265). The element linking rebirths is alone selected for attention in dependent origination because of its place in the Buddhist larger scheme of thought.

47. Contra David Kalupahana, *Causality: The Central Philosophy of Buddhism* (Honolulu: The University Press of Hawaii, 1975), pp. 54–59. Only when nescience is present does the formula encapsulate a causal chain covering the immediate segment of one's cycle of rebirths.

48. E.g., Jayatilleke, *Early Buddhist Theory of Knowledge* p. 449; Kalupahana, *Causality: The Central Philosophy of Buddhism,* p. 98.

49. Kenneth Inada, in his review of Kalupahana's *Causality* in *Philosophy East and West* 26 (1976): 344.

50. In general, mystics condemn any "magical" manipulation of sacred knowledge for selfish, mundane ends.

51. See Winston L. King, *A Thousand Lives Away* (Cambridge: Harvard University Press, 1964), pp. 98–99.

52. *Dhammapada* 1 claims that the *dhammā* are mind-made *(mano-māyā),* but this may refer to mental *dhammā* only. The orthodox Advaita position is a realism for the phenomenal realm, not one involving subjective creation.

Chapter 5. The Nature of Knowledge

1. S V.422 states that the divine eye, knowledge-that *(ñāna),* insight *(paññā),* wisdom *(vijjā),* and light all arise together when we realize that all is suffering. *Jñāna* / *ñana* for the Theravāda involves discrimination and thus is for the world of apparent multiplicity; for the enlightened state, this is the infusion of wisdom into perception or, in other words, an integration of wisdom and sensory or conceptual differentiation—i.e., the ability to utilize differentiations without projecting them.

2. M I.482, II.127. See Padmanabh S. Jaina, "On the *Sarvajñatva* (omniscience) of Mahāvīra and the Buddha," in L. Cousins et al., *Buddhist Studies in Honour of I. B. Horner* (Dordrecht, Holland: Reidel, 1974), pp. 76–83. By the time of the *Milindapañha,* literal omniscience was attributed to the Buddha, not merely knowledge of the world related to the origin and cessation of suffering (Mlp 74, 102, 106–7, 209–11).

3. For an expansion of this section, see Richard H. Jones, "Conceptualization and Experience in Mystical Knowledge," *Zygon* 18 (1983): 139–65.

4. *Viveka-cūdāmaṇi;* see Śaṁkara, *Shankara's Crest-Jewel of Discrimination,* p. 42.

5. Ibid., pp. 8, 110.

6. J. F. Staal, *Exploring Mysticism* (Berkeley: University of California Press, 1975), p. 25.

7. Cited in Aldous Huxley, *The Perennial Philosophy* (New York: Harper and Row, 1944), p. 132.

8. Mystics' accounts of their experiences will normally be neutral to the issue of whether the experience itself is structured by concepts or is only structured by concepts after the experience, since the accounts are always colored by the latter interpretation, just as an account of a red light is normally neutral to the issue of whether the light is red or in fact clear but covered with red glass. Only rarely do mystics see the problem, but when they do their claims favor the interpretation that the depth-mystical experience is free of all conceptual elements. See, e.g., the passage from Saint Teresa of Avila cited in William James, *The Varieties of Religious Experiences* (New York: New Library Books, 1958), pp. 313–14. Certainly one should not uncritically assert that the post-empiricist model of experience (that concepts pervade experiences) must be true of all experiences.

9. Clifford Geertz, *Islam Observed* (Chicago: The University of Chicago Press, 1968), pp. 17, 97.

10. Feyerabend, *Against Method,* p. 44.

11. Smart, *Doctrine and Argument in Indian Philosophy,* p. 78.

12. Staal, *Advaita and Neo-Platonism,* pp. 88–89, 158–60.

13. The apparent conflict of Judeo-Christian, Muslim, and Indian revelations also presents

doubts as to whether revelations are exempt from this problem even if the experiences are not of the same type.

14. Smart, *Doctrine and Argument in Indian Philosophy,* p. 86.

15. Lakatos, "Falsification and the Methodology of Scientific Research Programmes," in *Criticism and the Growth of Knowledge,* ed. Imre Lakatos and Alan Musgrave, pp. 91–195.

16. Paul W. Taylor, *Normative Discourse* (Englewood Cliffs: Prentice-Hall, 1961), p. 132.

17. Kalupahana, *Causality,* pp. 153f.

18. Jayatilleke, *Early Buddhist Theory of Knowledge,* p. 9. His position is qualified: a transempirical reality is implied in Buddhism (pp. 471–76).

19. Ibid., pp. 463–64, 331.

20. David Kalupahana, "A Buddhist Tract on Empiricism," *Philosophy East and West* 19 (1969): 65–67.

21. Jayatilleke, *Early Buddhist Theory of Knowledge,* p. 161.

22. Ibid., pp. 426, 462–63.

23. Kalupahana, *Causality,* pp. 49–50, 109, 130.

24. Cited in ibid., p. 130. The usual Theravādin claim is that the Buddhas know all of their past lives, while other enlightened persons do not.

25. Jayatilleke, *Early Buddhist Theory of Knowledge,* pp. 107–8.

26. Kalupahana, *Causality,* pp. 199–200.

27. See Ian Stevenson, *Twenty Cases Suggestive of Reincarnation* (Charlotesville: University Press of Virginia, 1974).

28. Kalupahana, *Causality,* p. 115; Jayatilleke, "Buddhism and the Scientific Revolution," p. 96.

29. If we could end death through medicine or whatever, and so achieve immortality, Theravāda Buddhism would not be logically undercut: the law of rebirth would still hold, but one antecedent condition (death) would not obtain. And we would still suffer disappointments from impermanence. *Nibbāna* would be falsified only by showing that no one can in principle attain the required state of desirelessness.

30. Some tribal cultures and some Western mystics influenced by Greek thought (especially that of Pythagoras and Plato) make previous and future lives, if not indeterminate cycles of rebirth, a minor topic. Desirelessness is discussed much more than ending rebirth, and is not usually seen as a means of ending any possible rebirth.

31. Smart, *Doctrine and Argument in Indian Philosophy,* p. 50.

32. Jayatilleke, *Early Buddhist Theory of Knowledge,* pp. 460, 96.

33. Ibid., pp. 441–42.

34. Jayatilleke, *Early Buddhist Theory of Knowledge,* p. 453.

35. Ibid., p. 97. For Śaṁkara, the self is the *only* verifiable reality.

36. E. A. Burtt, "The Buddhist Contribution to Philosophic Thought," in *Pathways of Buddhist Thought,* ed. M. O'C. Walshe (New York: Harper and Row, 1971), p. 83.

37. Even the Vedic Aryans appear to be considered empiricists. See Kalupahana, *Causality* pp. 1–5.

38. Harold K. Schilling emphasizes this in his *The New Consciousness in Science and Religion* (Philadelphia: United Church Press, 1973), pp. 301f.

39. Ronald Duncan and Miranda Weston-Smith, eds., *The Encyclopaedia of Ignorance* (New York: Pergamon Press, 1977), p. ix. The articles state what is accepted or under research at present, and reveal the eternal optimism of scientists (that what is important but not now known will be explained in the next decade).

40. Cited in Arthur Koestler, *The Act of Creation* (New York: Macmillan, 1964), p. 251.

41. Cited in Arthur Danto and Sidney Morgenbesser, eds., *Philosophy of Science* (New York: New American Library, 1960), p. 270.

42. Abraham H. Maslow, *The Psychology of Science* (Chicago: Henry Regnery, 1966), pp. 20–32.

43. Jacob Needleman, *A Sense of Cosmos* (New York: E. P. Dutton, 1976), pp. 15–16, 36.

44. To illustrate, names for parts of the brain mean such things as "inner room" (thalamus), "nose-brain" (rhinencephalon), and "interbrain" (diencephalon). They add nothing to our understanding and could positively mislead if taken literally, as in the case of "nose-brain."

45. Erwin R. Goodenough, *The Psychology of Religious Experience* (New York: Basic Books, 1965), chap. 1.

46. The mystical may be the reason the realm of becoming exists, but it cannot ultimately answer the question "why does *anything* exists?" any more than positing a creator god can. This maneuver merely pushes the problem back one step. Nor does the notion of a "self-caused" reality answer this question.

47. Non-theistic mystics do not speak of the *terrifying* qualify of the power of ultimate reality as often as some other types of religious persons do.

Chapter 6. Experiences

1. Aldous Huxley, *The Doors of Perception* (New York: Harper and Row, 1970), pp. 22–24.

2. James, *The Varities of Religious Experience* p. 298.

3. Tart, *States of Consciousness*, pp. 206–28.

4. Mario Bunge, *Intuition and Science* (Englewood Cliffs: Prentice-Hall, 1962).

5. See, e.g., Herbert A. Simon, "Does Scientific Discovery Have a Logic of Discovery?" *Philosophy of Science* 40 (1973): 471–80. His examples, though, are only of simple *laws*, not more imaginative *theories*.

6. It might also be added that while scientific training is in principle open to all, only those persons with rather high analytical abilities can in fact get beyond the basic ideas, or reproduce the delicate data.

7. E.g., Barbour, *Issues in Science and Religion*, p. 178.

8. That mystical experiences do not purport to have sensory objects does not make them necessarily hallucinatory, a hallucination being an experience of a supposedly public object that is not seen by others or does not even cohere with one's own sensory experiences. (Such hallucinations could also be replicated by following procedures strictly enough.)

9. A. J. Ayer, *The Central Questions of Philosophy* (New York: Holt, Rinehart and Winston, 1973), pp. 222–23.

10. A negative feature often attributed to mysticism is that mystics see only what they want to see and are easy to fool. The same has been said of scientists. See, e.g., Broad and Wade, *Betrayers of the Truth: Fraud and Deceit in the Halls of Science*, p. 107.

11. David Little and Sumner B. Twiss, Jr., *Comparative Religious Ethics* (New York: Harper and Row, 1978), p. 35; Israel Scheffler, *Science and Subjectivity* (New York: Bobbs-Merrill, 1967), pp. 2–3.

12. Alfred N. Whitehead, *Science and the Modern World* (New York: Macmillan, 1925), p. 3.

13. Popper, *Logic of Scientific Discovery*, p. 20.

14. Cited in Huston Smith, "Man's Western Way: An Essay on Reason and the Given," *Philosophy East and West* 22 (1972): 458.

15. Śaṁkara's claim is that the object alone determines knowledge (BSB I.1.2). Perception for Advaita, as for Aristotle and medieval Europeans, consists of the mind *(manas)* going out through the senses *(indriyā)* to an object. The mind is then modified *(vṛtti)* to "become" the form of the object. The understanding *(buddhi)* synthesizes the products of the mind into meaningful perceptions and ideas. The Theravādins' theory of perception is that sensation *(phassa)* results from three things coming together: the object of perception *(visaya)*, the sense-organ *(indriya)*, and discriminative awareness *(viññāṇa)*. An object is directly perceived, not inferred (as for the Sautrānikas). The real form of an object is seen when perception is free of all constructs *(nir-vikalpa-pratyākṣa)*, a form of nature-mysticism.

16. When mystical interests are part of mindfulness, these interests usually differ radically from scientific interests (if they do not directly conflict). For example, one Buddhist mindfulness-exercise is to observe a wound on the arm fester in order to see the "true nature" of the human body (its impermanence and inherent suffering). This is obviously not a scientific examination and will not lead to new medical knowledge. Nothing about meditative mindfulness could sustain scientific research.

Chapter 7. Language

1. For more general considerations of mystical utterances, see Richard H. Jones, "A Philosophical Analysis of Mystical Utterances," *Philosophy East and West* 29 (1979): 255–74.

2. Popper's falsification-criterion was originally advanced only to demarcate what was *scientifically interesting* from what was not, not as a criterion of meaningfulness of statements.

3. Jayatilleke, *Early Buddhist Theory of Knowledge,* pp. 471–76.

4. Thomas Luckmann, *The Invisible Religion* (New York: Macmillan, 1967). p. 54.

5. D. D. Lee, *Freedom and Culture* (Englewood Cliffs: Prentice-Hall, 1959); Berger and Luckmann, *The Social Construction of Reality* pp. 38–39; Tart, *States of Consciousness,* p. 45.

6. Ayer, *The Central Questions of Philosophy* p. 49.

7. Ibid.

8. Ravetz, *Scientific Knowledge and its Social Problems* p. 234.

9. Heisenberg, *Physics and Philosophy,* p. 168.

10. Heisenberg, *Physics and Beyond,* p. 208.

11. Ibid., pp. 129–30.

12. Quine, *The Ways of Paradox,* p. 219.

13. Arthur C. Danto, "Language and the Tao: Some Reflections on Ineffability," *Journal of Chinese Philosophy* 1 (1973): 45–55.

14. Burtt, *The Metaphysical Foundations,* p. 85.

15. N. R. Hanson, *Perception and Discovery: An Introduction to Scientific Inquiry* (San Francisco: Freeman, Cooper and Co., 1969), p. 25.

16. David Bohm, *Wholeness and the Implicate Order* (Boston: Routledge & Kegan Paul Ltd., 1981), chap. 2.

17. A broader philosophical problem—if all language is from the "dream," how can it refer to the "dreamer"?—involves the mirroring as only part of it.

18. See Bhikkhu Ñaṇananda, *Concept and Reality in Early Buddhist Thought* (Kandy, Sri Lanka: Buddhist Publication Society, 1971).

19. Later the idea of two "levels of truth"—the conventional and the ultimate—developed, but this idea is not present in the Pāli canon. It is sometimes asserted that science too has a two-level view of truth—the commonsensical and the scientific points of view. (See, e.g., Jayatilleke, "The Buddhist Conception of Truth" in *Pathways of Buddhist Thought,* ed. M. O'C. Walshe, p. 80). The commonsensical table is convenient to talk about, but the atomic table is the reality. If only statements about the latter are considered to be true, this would presume a reductive metaphysics. But while it makes sense to speak of different contexts for different claims ("solidity" as an everyday concept versus as an atomic concept), still there are difficulties in a strict comparison between science and the Indian idea. Primarily, the problem is that there are many different scientific levels—astronomical, geological, biological, chemical, physical—unless we accept only the lowest level of organization as the only legitimate one. More relevant to the Indian theory of levels of *truths* (rather than the ontological nature- and depth-mystical levels) are theories that are convenient on the everyday level but are ultimately unsatisfactory (e.g., Newtonian physics), and most relevant are conventions used within theories.

20. Ñaṇananda, *Concept and Reality,* p. 35.

21. Even if nothing can be said about some "ultimate reality" beyond the *dhammā,* still

certain "ultimate" truths (such as that one) can be stated and are never corrected or contradicted by further experience.

22. Danto, "Language and the Tao," pp. 51–52; Scheffler, *Science and Subjectivity,* pp. 27–28.

23. Heisenberg, *Physics and Beyond,* p. 210.

24. See, for example, Max Black, "Metaphor" in his *Models and Metaphors* (Ithaca: Cornell University Press, 1962) or Paul Henle, "Metaphor," in *Language, Thought, and Culture* (Ann Arbor: Michigan University Press, 1958), pp. 173–95.

25. Hesse, *Models and Analogies in Science,* pp. 150–56.

26. Both foci (or even more) are often intended.

27. Berger and Luckmann, *Social Construction of Reality,* pp. 40, 95. According to them, religion and science are two of several symbolic representations of reality.

28. Heisenberg, *Physics and Beyond,* p. 210.

29. Cited in Schilling, *The New Consciousness in Science and Religion,* p. 81.

30. Hesse, *Models and Analogies in Science.* She goes on to argue that theories as a whole are analogies; see "Theory as Analogy" in her *The Structure of Scientific Inference* (Berkeley/Los Angeles: University of California Press, 1974), pp. 197–222.

31. Stephen Toulmin, *The Philosophy of Science* (New York: Harper and Row, 1953), p. 13.

32. Carl G. Hempel, *Philosophy and Natural Science* (Englewood Cliffs: Prentice-Hall, 1966), p. 70.

33. See Hanson, *Patterns of Discovery,* pp. 50–69.

34. This introduces the problem that the original context giving meaning to a metaphor may be lost; subsequent understanding of the metaphor would then be altered.

35. Some mystical traditions do have secret languages, i.e., codes to protect the sacred knowledge from the general public (and vice versa). But oral commentaries decode these languages for the initiates.

36. If mystics were to invent an entirely new language (vocabulary and grammar), still the problems of the mirror-theory of language would remain. Perhaps this is why mystics do not bother to attempt it.

37. It may be that in a metaphor the *meaning* is also changed somewhat by association with something new. The meaning of "one" or "real" would have to change if an analogy of proportionality is involved between ordinary and mystical uses. The problem: if the meaning changes, how are we to understand the utterances?

38. Cited in David L. Gosling, *Science and Religion in India* (Madras: The Christian Literature Society, 1976), p. 158.

39. See, e.g., King, *A Thousand Lives Away,* pp. 129–30.

40. Imre Lakatos, *Proofs and Refutations: The Logic of Mathematical Discovery,* ed. John Worrall and Elie Zahar, (New York: Cambridge University Press, 1976). He concedes that he is looking at the "near empirical" parts of mathematics, i.e., the more informal parts. His conclusions reflect those of this teacher (Popper) on science: mathematics is a matter of conjectures and refutations. But also see Morris Kline, *Mathematics: The Loss of Certainty* (New York: Oxford University Press, 1980), pp. 314–320, on the nature of mathematical proof. Even in mathematics, "the proofs of one generation are the fallacies of the next." (p. 318)

41. Even logic has conventions; e.g., logicians find it convenient to treat contradictions (the conjunction of *a* and not-*a*) as *false* claims, not *meaningless* utterances. For a discussion of the circularity in the justification of deduction, see Susan Haack, "The Justification of Deduction," *Mind,* n.s., 85 (1976): 112–19.

42. Any altered states of consciousness available for the manipulation of concepts would differ from mystical ones for this reason.

43. Willard V. O. Quine, *Word and Object* (Cambridge: MIT Press, 1960), chap. 1.

44. Whether mathematics is merely our convenient system or is the "language of reality" has remained undecided since the time of the ancient Greeks. Many traditional peoples considered it, like all language, a power, with the numbers as creative principles. In fact, a version of

the Pythagorean idea that "the nature of things is number" was instrumental in the rise of modern science with Galileo's claim that what is real is only what is quantifiable.

45. Werner Heisenberg, *Across Frontiers* (New York: Harper and Row, 1974), p. 146.

Chapter 8. Possible Relationships between Scientific and Mystical Claims

1. Kenneth R. Pelletier, *Toward a Science of Consciousness* (New York: Dell Publishing, 1978), p. 142 (on biofeedback).

2. For a discussion of some of the issues involved in comparing psychology and Eastern religious thought, see Richard H. Jones, "Jung and Eastern Religious Traditions," *Religion* 9 (1979): 141–56.

3. Erwin Schrödinger, *My View of the World,* trans. Cecily Hastings (Cambridge: Cambridge University Press, 1964).

4. As Jeremy Bernstein says in his "Popular Science," *New Yorker* 55 (October 8, 1979): 169–70, tying religion to contemporary physics is a sure route to its obsolescence. Many of the scientific theories relied upon by such people as Capra and Zukav (e.g., bootstrap theory) have lost their popularity in the last decade among most physicists. In addition, the conflicts within the "new physics" should not be overlooked. For example, in bootstrap theory, unlike in relativity, there is no continuous space-time or fields; and the apparent indeterminism of quantum events has generated various competing explanations (the Copenhagen interpretation, Wigner's theory of consciousness as the hidden variable, and the multiple-world theory). Such theories cannot be amalgamated indiscriminately.

5. Granville C. Henry, Jr., *Logos: Mathematics and Christian Theology* (Lewisburg: Bucknell University Press, 1976), p. 27.

6. Leon Cooper, "Do Values Play a Role in Science?" *Brown Alumni Monthly,* May/June 1977, p. 21.

7. The value of comparisons also is hurt by comparers' extensive reliance upon popular presentations of Eastern thought (e.g., the later works of D. T. Suzuki), which have been heavily influenced by Western psychology and popular notions connected to science. The danger of circularity in such circumstances is very real.

8. Gary Zukav, *The Dancing Wu Li Masters: An Overview of the New Physics* (New York: William Morrow, 1979), p. 104.

9. Geoffrey Chew, "Hadron Bootstrap: Triumph or Frustration?" *Physics Today* 23 (October, 1970): 28.

10. Barbour, *Issues in Science and Religion* p. 290.

11. Heisenberg, *Physics and Philosophy,* p. 199.

12. See Lawrence LeShan, *The Medium, the Mystic and the Physicist* (New York: Ballantine Books, 1966).

13. Fritjof Capra, *The Tao of Physics* (Boulder: Shambhala Publications, 1975), p. 114.

14. For various positions advocated by scientists themselves see Gerald Holton, "Notes on the Religious Orientation of Scientists," in *Science Ponders Religion,* ed. Harlow Shapley (New York: Appleton-Century-Crofts, 1960), pp. 52–64, and Gosling, *Science and Religion in India.*

15. Goodenough, *The Psychology of Religious Experience,* chap. 1.

16. Ralph W. Burhoe's "translation" of "God" as "the ultimate necessities of laws and boundary conditions imposed by nature" (p. 195) and of the phrase "unless the Lord builds the house" as "unless the builders operate in accord with what the nature of things will permit" (p. 172) suggests such a position (*Science and Human Values in the 21st Century,* [Philadelphia: Westminster Press, 1971]).

17. See Barbour, *Issues in Science and Religion,* pp. 428–30.

18. Creationism also involves trying to find scientific support for specific theories (pre-

sented by a literal reading of the Bible) held to be true independent of scientific results. All science has a metaphysical framework held independently of specific theories; but creationism differs from science in having conclusions regarding specific theories, rather than the broader framework, in advance of research.

19. Representative discussions: William H. Austin, "Complementarity and Theological Paradox," *Zygon* 2 (1967): 365–81, and James L. Park, "Complementarity without Paradox," *Zygon* 2 (1967): 382–88.

20. Heisenberg, *Physics and Philosophy,* pp. 21, 179. Niels Bohr, *Atomic Physics and Human Knowledge* (New York: Science Editions, Inc., 1958), pp. 20–21, 74–77, and 100–101.

21. J. Robert Oppenheimer, *Science and the Common Understanding* (New York: Simon and Schuster, 1958), p. 69.

22. If we focus upon *understanding* rather than modes of awareness, there would have to be one demonstrably correct interpretation of mystical experiences that complements a unique scientific construal of the world. To mention only one problem, at present quantum theory and relativity "directly contradict each other" (Bohm, *Wholeness and the Implicate Order,* p. 176).

23. It should be added that even some scientists (including Einstein) claim not to understand the scientific application of the idea of "complementarity."

24. F. S. C. Northrop's introduction to Heisenberg, *Physics and Philosophy,* p. 21.

25. Raymond B. Blakney, trans., *Meister Eckhart* (New York: Harper and Row, 1941), pp. 213, 85.

26. Some mystics may deny even this. Within Islam, the Ash'arite doctrine is that God is the only actor—there is not autonomous natural order, and hence no real causal order among appearances.

27. A negative judgment on the part of scientists towards mysticism is not necessary: it can be conceded that there is more to reality than science discloses. Only the claim that mysticism along gives knowledge must be denied.

28. Rarely, passages suggesting that scientific knowledge is given in mysticism appear. For example, the *Yoga-sūtra* says "inner control" *(saṃyama)* on the navel gives knowledge of the bodily system. But this knowledge is of the yogic physiology of centers *(cakras)* and channels, i.e., old knowledge is internalized but no new knowledge is revealed.

Chapter 9. Cosmology, Physics, and Mysticism

1. Traditional societies also have their own competing "cosmologies" in a more scientific sense in that such cosmologies are not closely tied to ways of life.

2. See R. Penrose, "Singularities and Time-Assymmetries," *General Relativity: An Einstein Centenary Survey,* ed. S. W. Hawkins and W. Israel (New York: Cambridge University Press, 1979), pp. 630–34; and Paul Davies, *Other Worlds* (New York: Simon and Schuster, 1980), pp. 128–82.

3. Many would argue that the Genesis account is compatible with any scientific cosmology (e.g., the steady-state theory of the continuous creation of matter). This would involve adjusting belief to a theory of God as a "ground of being" maintaining reality; the accounts in Genesis would need to be seen as symbolic of the univere's dependence.

4. Jastrow, *God and the Astronomers,* p. 14.

5. Sāṁkhya, in fact, is felt to be the classical Indian school most compatible with science (especially with biology). See D. M. Bose, ed., *A Concise History of Science in India* (New Delhi: Indian National Science Academy, 1971), p. 25. Sāṁkhya sounds scientific with its talk of "evolution" *(sṛṣṭi).* But it is an "emanation" of matter *(prakṛti)* from unmanifest forms to coarse objects—a metaphysical "evolution"—that is being discussed, not evolution in the biological or cosmological sense of a process within one phenomenal level.

6. R. F. Gombrich, "Ancient Indian Cosmology," in *Ancient Cosmologies,* ed. Carmen Blacker and Michael Loewe (London: George Allen and Unwin, 1975), p. 110.

7. L. de La Vallée Poussin, "Ages of the World (Buddhist)," *Encyclopedia of Religion and Ethics,* 1: 188.

8. King, *A Thousand Lives Away* p. 107.

9. Of the many medieval Indian mythological calculations of a world-age, the standard one (approximately 4.3 billion years long) is huge by a historical framework, but short by an astronomical one. The grand scale is in common with science, but the Indian version is merely a correlate of other huge figures (e.g., the number of gods).

10. Jayatilleke, "Buddhism and the Scientific Revolution," p. 93.

11. King, *A Thousand Lives Away,* p. 98.

12. Furthermore, the "night of Brahma" (when matter is not unrolled) lasts as long as creation, unlike in science, where time cannot be measured in the absence of movement.

13. Buddhist and other Indian texts depict different ends to each world-age (by fire, water, or wind), depending upon the type of evil predominating in the age. The new universe emerges out of primal waters.

14. Cited in Gosling, *Science and Religion in India,* p. 140.

15. From the relativity of motion, we cannot tell whether the earth or the sun is moving; but coupling this situation with other scientific theories (e.g., from mechanics, the heavier object controls the lighter one), we can conclude that if the earth is stationary, our conception of the universe is radically erroneous. Thus, not all of physics is relativized.

16. Heisenberg, *Physics and Philosophy,* p. 80.

17. If we assume that the four basic physical forces can be unified, still the mystical "forces" (behind paranormal connections and the "power" of being) are not covered. The power of being is conserved only in that it always remains identical.

18. Quarks are not discussed much by writers comparing Western science and Eastern thought. To most commentators, the search for fundamental, indivisible, unchanging building-blocks of matter is an old-fashioned and outdated approach in science. There are indeed problems: quarks were initially introduced to bring simplicity to this realm, but they have become so complex that some physicists have likened them to Ptolemaic epicycles and have questioned the whole quest for elementary "pieces" of matter.

19. Heisenberg, *Physics and Philosophy,* p. 202.

20. Capra, *The Tao of Physics,* p. 215. Associating voidness with relativistic space-time again has the problem that it was never tied especially to *akāśa* but directly to all experienced reality.

21. See Bohm, *Wholeness and the Implicate Order,* for one physicist's philosophical discussion of holisms.

22. Bohr contended that the closed processes in quantum physics are not directly analogous to biological functions (*Atomic Physics and Human Knowledge,* p. 76), e.g., there is no adaptation to the environment (p. 100).

23. Sometimes paranormal knowledge of the future and past is considered to be necessarily unmechanical. But if, as was suggested in a previous footnote, this involves "reading" the kammic seeds present at the moment of experience, this allegation is wrong.

24. See Richard H. Jones, "The Nature and Function of Nāgārjuna's Arguments," *Philosophy East and West* 28 (1978): 485–502.

25. The one reality underlying everything is "in" everything. Thus the reality in one fragment of reality is in each other fragment and vice versa, i.e., only one reality is involved. Nescience is taking each nexus to be self-contained.

26. D. T. Suzuki, *On Indian Mahayana Buddhism,* edited by Edward Conze (New York: Harper and Row, 1968), p. 167.

27. Garma C. C. Chang, *The Buddhist Teaching of Totality* (University Park: The Pennsylvania State University Press, 1971), p. 124.

28. Ibid., p. 123. "Reflects" here also means "related to" or "dependent upon" (p. 125).

29. Bohm on "formative causes" (*Wholeness and the Implicate Order,* pp. 12–14) is relevant to this question.

30. Suzuki, *On Indian Mahayana Buddhism*, p. 149.

31. It must be noted that this theory is a rival to the quark theory, i.e., it is a different approach to identifying building-blocks of matter. Currently the quark theory, with all its problems, is much more popular. For Chew's defense of bootstrap, see *The Analytic S-Matrix* (New York: W. A. Benjamin, 1966).

32. Bohn, *Wholeness and the Implicate Order*, chap. 6. Bohm uses "holomovement" to emphasize the dynamic nature of reality, unlike the static nature of holograms.

33. Suzuki, *On Indian Mahayana Buddhism*, p. 168.

34. Since the depth-mystical is utterly structureless, there is no reason to suggest, as Bohm does, that mystics experience an implicate order underlying all structure in phenomena, since such an order must itself be structured to account for other structures.

35. Sāṁkhya-Yoga is a dualism that is the opposite of any holism: awareness and matter are isolated from each other.

36. Heisenberg, *Physics and Philosophy*, p. 58.

37. Eugene P. Wigner, *Symmetries and Reflections* (Bloomington: Indiana University Press, 1967). See also C. T. K. Chari, "Quantum Physics and Concepts of Consciousness," *Indian Philosophical Annual* 10 (1976): 50–56.

38. Heisenberg, *Physics and Beyond*, pp. 129–30.

39. Ian G. Barbour, *Myths, Models, and Paradigms* (New York: Harper and Row, 1974), p. 161.

40. Some concepts (e.g., conservation laws) are believed to be applicable in each of these realms.

41. Zukav, *The Dancing Wu Li Masters*, p. 101.

42. Cited in Ernest Nagel, *Logic Without Metaphysics* (New York: The Free Press, 1958), p. 322; cf. Oppenheimer, *Science and the Common Understanding*, pp. 74–76.

43. Schrödinger, *My View of the World*, pp. vii–viii.

44. Watts, *Psychotherapy East and West*, p. 82.

45. From this point of view, Bohm's idea of an implicate order—an unmanifest, self-ordering structure—would be merely another level within the "dream."

46. Kenneth R. Pelletier, *Toward a Science of Consciousness*, p. 253.

47. Arthur J. Deikman, "Deautomatization and the Mystic Experience," reprinted in *The Nature of Human Consciousness*, ed. Robert E. Ornstein (San Francisco: W. H. Freeman, 1973), p. 233.

48. F. S. C. Northrop, *The Logic of the Sciences and the Humanities* (New York: Macmillan, 1947), p. 48.

49. All numbers in the text of this section are page references to Northrop's *The Meeting of East and West* (New York: Colliers, 1966).

50. Pp. 340 and 377; also *The Logic of the Sciences and the Humanities*, p. 387.

51. Ibid., p. 378.

52. Ibid., p. 48.

53. Ibid., p. 82.

54. Ibid., p. 195.

55. P. 447; Northrop, *The Logic of the Sciences and the Humanities*, p. 63.

56. Ibid., p. 100.

57. Page references here will be to Capra's *The Tao of Physics*, 2d ed. (Boulder: Shambhala Publications, 1983) (the second edition updates the physics but does not change the material on Eastern thought or the comparisons). Seeing Śiva as the "dancing energy" of a nature-mystical experience may seem natural, even though this god is never so described nor do his dances of joy (over his power) and destruction (at the end of the world-age) have anything whatsoever in common with creation or the destruction in the atomic realm.

58. Fritjof Capra, "Bootstrap and Buddhism," *American Journal of Physics* 42 (1974): 19; "The Tao of Physics: Reflections on the Cosmic Dance," *Saturday Review*, December 10, 1977, p. 21.

59. Capra, "The Tao of Physics: Reflections on the Cosmic Dance," p. 21. The language in

this paragraph was not changed in the revised edition, but in other recent writings Capra does not speak of "confirmation" but of physics as "mirroring" mysticism.

60. Capra, "Bootstrap and Buddhism," p. 19.

61. Capra, "The Tao of Physics: Reflections on the Cosmic Dance," pp. 21–22; *The Tao of Physics,* p. 12.

62. The traditional particle-approach (represented by the quest for quarks) currently guides most research. But Capra and Chew still defend the bootstrap approach and are expanding it to cover other particles on the same level. See Capra, "Bootstrap Theory of Particles," *ReVision* 4 (1981): 88–91.

63. Capra, "Bootstrap and Buddhism," p. 19.

64. Also "The Tao of Physics: Reflections on the Cosmic Dance," p. 28.

65. Toulmin, "Contemporary Scientific Mythology."

66. Also "The Tao of Physics: Reflections on the Cosmic Dance," pp. 23–24, 28.

67. In one conversation, Capra said that while physicists explore levels of matter and mystics explore levels of the mind, what they have in common is that the levels explored "lie beyond ordinary sensory perception." (*The Holographic Paradigm and Other Paradoxes,* ed. Ken Wilbur, [Boulder: Shambhala Publications, 1982], p. 231.) Ignoring the fact that mystical claims are about more than mind alone, still Capra's claim reduces to a not very illuminating point of abstract commonality. The similarities between "patterns and principles of organization" that Capra sees in submicroscopic physics and mysticism are also of the same degree. (Capra's assertion in the same conversation (ibid., pp. 231–32) that mystics do not make claims about the ordinary realm is hard to support.)

68. Also Capra, "The Tao of Physics: "Reflections on the Cosmic Dance," p. 22.

69. Ibid., pp. 22, 28.

70. Capra, "Bootstrap and Buddhism," p. 16.

71. Page references in the rest of this section are to Gary Zukav, *The Dancing Wu Li Masters* (New York: William Morrow, 1979).

72. P. 254. Zukav's characterization of Buddhism here also includes reference to the past and future as being illusions. But Buddhist belief in *kamma* (which involves past acts and future fruits) and rebirth should be sufficient to counter this claim. Only in the extreme state of mindfulness, in which all concepts are eliminated, is time ignored.

73. This "public case" method is generally considered in Japan to be inefficient in inducing enlightenment; most often disciples have to work through a long list of such cases before, if ever, a dramatic insight into the *general* nature of concepts occurs (i.e., that reality as it really is cannot mirror the differentiated concepts projected upon it). Perhaps this is because, unlike other meditative techniques, the "public case" method increases the working of the analytical power of mind in an attempt to suspend it.

Chapter 10. A Reconciliation of Science and Mysticism

1. Even the statement of this problem is only one philosophical position.

2. For classical statements on both sides of this issue, see W. K. Clifford, "The Ethics of Belief," and William James, "The Will to Believe," reprinted in *Religion from Tolstoy to Camus,* ed. Water Kaufmann (New York: Harper and Row, 1961), pp. 201–38.

3. Ludwig Wittgenstein, *On Certainty;* Norman Malcolm, "On the Groundlessness of Belief" in *Reason and Religion,* ed. Stuart C. Brown (Ithaca: Cornell University Press, 1977), pp. 143–57.

4. Heisenberg, *Physics and Philosophy,* p. 205.

5. Hanson asks why we should accept some particular scientific theories (although not scientific theories *per se*) and answers: "Because the world as we now know it becomes intelligible by supposing these things to be the case. What better reason for saying that they are

the case?" (*Patterns of Discovery*, p. 134.) This shows the difficulty of trying to come up with some justification of what we normally take to be obvious.

6. Needleman, *A Sense of Cosmos*, pp. 13–14.

Appendix: Concerning the Possible Philosophical Significance of Scientific Studies of Mysticism

1. The reverse situation—mystical studies of science—does not occur because of the mystic's general lack of interest in science, as discussed in chapter 8.

2. Herbert Benson, *The Relaxation Response* (New York: William Morrow, 1975), p. 114 (italics added); cf. pp. 75–78. It is not clear whether he means only that science has shown a physiological effect of meditation. Of course "age-old wisdom" is not that meditation lowers blood-pressure or has any other such effect. This ambiguity points to the usual lack of precise thought in physiological studies concerning their philosophical implications.

3. Tart, *States of Consciousness*, p. 207. For some physiologists, consciousness is not necessarily unreal for being unobservable: it can still produce an effect. Behavioralists hold another view.

4. Robert E. Ornstein, *The Psychology of Consciousness* (New York: Penguin, 1975), pp. 25, 218.

5. Ornstein, *The Mind Field* pp. 124–25.

6. Benson, *The Relaxation Response*, p. 19. In his more recent book, *Beyond the Relaxation Response* (New York: Times Books, 1984), Benson explores the physiological effect of the relaxation response combined with the "faith factor" (meditation and a deeply held set of philosophical or religious convictions).

7. Cited in Staal, *Exploring Mysticism*, p. 104. This may mean only that we need more research to determine what is actually important here.

8. The *learned* aspect is essential to meditation too. Progress has recently been made regarding how pain-killing placebos work. But still, how sugar tablets can touch off the production of endorphin only in certain circumstances will need reference to the ultimate role of consciousness.

9. Tart, *States of Consciousness*, pp. 152–53.

10. Meditative positions and practices may only prepare the practitioner for enlightenment and not actually produce it. But those who practice meditation do increase the chance of enlightenment (whether through the techniques themselves or by the fact that those who are predisposed to practice meditation are also those who are predisposed to have mystical experiences). Correlations may be possible here.

11. Any explanation would also have to account for the dry "dark nights of the soul" in which no mystical experiences occur.

12. Benson, *The Relaxation Response*, p. 115.

13. Ayer, *The Central Problems of Philosophy*, p. 130. If there is a *causal* order between the conscious and the physical, one correlate will need to be given greater weight, since only a correlation appears.

14. Any correlation here will have to take into account that there is other activity allegedly dominated by the right hemisphere (e.g., intuitions in science).

15. William Alston, "Psychoanalytic Theory and Theistic Belief," in *Faith and the Philosophers*, ed. John Hick (New York: St. Martin's Press, 1966), p. 90.

16. Concerning whether specifying the psychological origin of mysticism disproves it, Ayer says that a psychological reduction is unreasonable if mystical experiences satisfy the criteria of objectivity (i.e., if a great number of people have the experience and each gives a similar description). See *The Central Problems of Philosophy*, pp. 222–23.

17. A point made by Norman Malcolm ("Is it a Religious Belief that 'God Exists?'" in *Faith*

and the Philosophers, ed. John Hick, pp. 103–10) and by William J. Wainwright ("Natural Explanations and Religious Experience," *Ratio* 15 [1973]: 98–101) in discussing Alston's article.

18. James, *The Varieties of Religious Experience,* pp. 29–30.

19. Wilder Penfield, *The Mystery of the Mind* (Princeton: Princeton University Press, 1975), p. 77.

20. James, *The Varieties of Religious Experience,* p. 30. Alston notes James's position with respect to a neurotic temperament as perhaps being necessary for inspiration from God, and quite literally does not know what to say ("Psychoanalytic Theory and Theistic Belief," p. 98).

21. Claudio Naranjo and Robert E. Ornstein, *On the Psychology of Meditation* (New York: Viking Press, 1971), p. 196.

22. Arthur Eddington, *The Nature of the Physical World,* pp. 23f.

23. Pelletier, *Toward a Science of Consciousness,* pp. 103–4. Nor is the specialization absolute compartmentalization: the right hemisphere processes words too (as word-pictures such as metaphors) (pp. 101–2).

24. Staal comes to the conclusion that physiological studies do not throw light on the larger issue of whether knowledge is obtained by means of meditation; they only establish that "meditation has definite effects on the body, which remain unexplained and are probably side effects of little importance" (*Exploring Mysticism,* p. 110).

25. Biofeedback (the electronic monitoring of physiological data enabling self-control of various bodily mechanisms) is a form of a behavioralist technology that seems effective in producing calm, if not in producing many mystical enlightenment-experiences.

26. It should be added that any possible verification of these various paranormal powers contributes nothing to the independent epistemological question of whether enlightenment-claims are themselves correct.

27. E. E. Evans-Pritchard, *Theories of Primitive Religion* (London: Oxford University Press, 1965), p. 111. The functionalist approach will not be discussed here. For a recent discussion of ths approach, see Robert N. McCauley and E. Thomas Lawson, "Functionalism Reconsidered," *History of Religions* 23 (1984): 373–81.

28. The contrast between explanatory and descriptive explanations can be illustrated by two remarks. First from Mircea Eliade: "The division of the village into four sections . . . *corresponds* to the division of the universe into four sections" (*The Sacred and the Profane* [New York: Harcourt, Brace & World, Inc., 1959], p. 45; emphasis added). Durkheim comments upon a similar situation but causally interpreted: "There are societies in Australia and North America where space is conceived in the form of an immense circle, *because* the camp has a circular form" (*The Elementary Forms of the Religious Life* [New York: The Free Press, 1965], p. 24; emphasis added). Durkheim does not explain how he knows that the causal order is not reversed, or how this claim could be tested.

29. Delving into such explanations also raises the possibility that different socioscientific accounts will *conflict* over the ultimate nature of a personal or social phenomenon. Different physiological accounts are usually seen as paralleling each other, not conflicting; neurology need not be denied in order to affirm accounts of muscle-mechanisms. Each physiological perspective covers any given phenomenon completely from its point of view without exhausting the total reality involved. Conflicts occur only if we have a metaphysics of the ultimate factors involved in a reality.

30. Robert N. Bellah, "Confessions of a Former Establishment Fundamentalist," *Bullentin of the Council on the Study of Religion* 1 (1970): 3–6. Also see his "Christianity and Symbolic Realism," *Journal for the Scientific Study of Religion* 9 (1970): 89–99, and Peter Berger, "Some Second Thoughts on Substantive versus Functional Definitions of Religion," *Journal for the Scientific Study of Religion* 13 (1974): 125–33.

31. Edmund Leach, *Political Systems of Highland Burma* (London: Bell, 1954), p. 182; italics added.

32. Before these socioscientific theorists, solar mythologists saw the same myths as symbols of *natural* phenomena such as the sun or rain.

33. Contra Hans H. Penner and Edward A. Yonan, "Is a Science of Religion Possible?"

Journal of Religion 52 (1972): 107–33, if this socioscientific fundamentalism at least is meant. See John Y. Fenton, "Reductionism in the Study of Religions," *Soundings* 53 (1970): 61–76.

34. A reply to Alasdair MacIntyre, "Is Understanding Religion Compatible with Believing?" in *Faith and the Philosophers,* ed. John Hick, pp. 115–33.

35. To treat the sacred as a real factor that cannot be grasped by nonreligious approaches is not reductive, contra Penner and Yolton, "Is a Science of Religion Possible?" and also Fenton, "Reductionism in the Study of Religions." Concepts which focus attention upon limited aspects of a subject are not as such reductive; reductionism is involved only when the *reality* of other aspects is questioned.

36. Melford E. Spiro, "Religion: Problems of Definition and Explanation," in *Anthropological Approaches to the Study of Religion,* ed. Michael Banton (London: Tavistock, 1966), p. 123; also see his *Buddhism and Society* (New York: Harper and Row, 1970), pp. 26–28.

37. Leach, *Political Systems of Highland Burma* p. 14.

38. The reason most often cited for ignoring the participants' own explanations of their practices is that different members of the same society give different explanations (or no explanation), and that different societies with *prima facie* somewhat similar rituals give radically different explanations. This compares within the physiological study of meditation, with dismissing all various religious understanding as irrelevant to the real effects involved. This obviously involves a theory itself and is not deducible from the scientific findings. In the case of different explanations of, say, the nature of the sun within even one culture, no one would conclude that physiological or social phenomena connected to observations are alone real, not the sun; nor that there might not be one legitimate view of the nature of the sun.

39. See Ian Hamnett, "Sociology of Religion and Sociology of Error," *Religion* 3 (1973): 1–12.

40. For a delineation of possible relationships between psychological and mystical "health," see Jones, "Jung and Eastern Religious Traditions," pp. 149–53.

41. For example, see Evelyn Underhill, *Practical Mysticism* (New York: E. P. Dutton, 1915), pp. viii–xi. Psychological crises may break the hold of our everyday expectations and thus supply conditions necessary for enlightenment-experiences; but there are other means of supplying these conditions, and not all such crises lead to these experiences. Thus these crises are not necessary nor sufficient for the occurrence of mystical enlightenment; nor, for the same reasoning, are social pressures.

42. Maslow, *The Psychology of Science,* pp. 20–32.

43. The claim that "all reasons are just rationalizations determined by underlying socio-psychological causes" is itself a theory; by its own terms, advancing reasons for it would be self-defeating. Exempting one's own reasons—i.e., treating only the participant's or one's opponents' reasons as rationalizations—would obviously require an argument (reasons) that does not beg the important issues. If an exemption to the claim is advanced, we once again would be trying to find the correct theory. The *reductio ad absurdum* of the basic position would be to find, for example, psychological causes of why Freud advanced a thesis used as a reason to invalidate giving reasons; see Stephen Toulmin, *Knowing and Acting* (New York: Macmillan, 1976), pp. 29–43, and John Greene, *Science, Ideology and Worldview* (Berkeley/Los Angeles: University of California Press, 1981), pp. 1–2. The only alternative interpretation of the position that all reasons are rationalizations of the real underlying causes is that it cannot be defended or criticized rationally; any reasons offered for holding or rejecting it would in turn be only rationalizations. The position, like all theories, would be immune from criticism, but could not be used as a reason against any other position; it would be completely isolated, and any discussion would be ultimately valueless.

44. Ludwig Wittgenstein, *On Certainty,* paras. 166, 559.

45. The difference between the physiological and psychosocial situations may be because most human action is purposive (end-oriented) rather than just mechanical. The meaning of actions, not just observables, is necessary to understand them. For this, intentions and goals related to needs and desires are important, and may be lawful (correlatable).

Select Bibliography

1. Western Science and Eastern Thought

Bernstein, Jeremy. "A Cosmic Flow." *American Scholar* 48 (1979): 6–9.

———. "Popular Science." *New Yorker* 55 (October 8, 1979): 169–70.

Bohm, David. *Wholeness and the Implicate Order.* Boston: Routledge & Kegan Paul Ltd., 1981.

——. "The Physicist and the Mystic—Is a Dialogue Between Them Possible?" In *The Holographic Paradigm, and Other Paradoxes,* edited by Ken Wilbur. Boulder: Shambhala Publications, 1982.

Burkhardt, Titus. "Cosmology and Modern Science." Reprinted in *The Sword of Gnosis: Metaphysics, Cosmology, Tradition, Symbolism,* edited by Jacob Needleman. Baltimore: Penguin Books, 1974. Pp. 122–78.

Capra, Fritjof. "Bootstrap and Buddhism." *American Journal of Physics* 42 (1974): 15–19.

———. *The Tao of Physics.* Boulder: Shambhala Publications, 1975; 2d ed., 1983.

———. "The Tao of Physics: Reflections on the Cosmic Dance," *Saturday Review* (December 10, 1977), pp. 21–23, 28.

———. "*The Tao of Physics* Revisited: A Conversation with Fritjof Capra." In Ken Wilbur (ed.), *The Holographic Paradigm,* edited by Ken Wilbur. Boulder: Shambhala Publications, 1982.

Chari, C. T. K. "Quantum Physics and East-West Rapprochement." *Philosophy East and West* 5 (1955): 61–67.

———. "Quantum Mechanics and Concepts of Consciousness," *Indian Philosophical Annual* 10 (1976): 50–56.

Dahlke, Paul. *Buddhism and Science.* London: Macmillan, 1913.

Esbenshade, Donald H., Jr. "Relating Mystical Concepts to Those of Physics: Some Concerns." *American Journal of Physics* 50 (1982): 224–28.

Grof, Stanislav, ed., *Ancient Wisdom and Modern Science.* Albany: State University of New York Press, 1984.

Harrison, David. "What You See Is What You Get!" *American Journal of Physics* 47 (1979): 576–82.

———. "Teaching *The Tao of Physics.*" *American Journal of Physics* 47 (1979): 779–83.

Jayatilleke, K. N. "Buddhism and the Scientific Revolution." In M. O'C. Walshe (selector), *Pathways of Buddhist Thought.* New York: Harper & Row, 1971.

Jones, Richard H. "Mysticism and Science: Against Needham on Taoism." *Journal of Chinese Philosophy* 8 (1981): 245–66.

King, Winston L. *A Thousand Lives Away: Buddhism in Contemporary Burma.* Cambridge: Harvard University Press, 1964. Pp. 85–146.

LeShan, Lawrence. *The Medium, the Mystic, and the Physicist.* New York: Random House, 1966.

———. *Alternative Realities.* New York: Ballantine Books, 1976.

Nasr, Seyyed Hossein. *Western Science and Asian Cultures.* New Delhi: Indian Council For Cultural Relations, 1976.

———. *Knowledge and the Sacred.* New York: The Crossroad Publishing Company, 1981.

———. "The Role of the Traditional Sciences in the Encounter of Religion and Science—An Oriental Perspective." *Religious Studies* 20 (1984): 519–41.

Needleman, Jacob. *A Sense of Cosmos.* Garden City: Doubleday, 1975.

Northrop, F. S. C. "The Complementary Emphases of Eastern Intuitive and Western Scientific Philosophy." In *Philosophy East and West,* edited by Charles A. Moore. Princeton: Princeton University Press, 1944.

———. *The Meeting of East and West.* New York: Macmillan, 1946. Pp. 294–404, 440–54.

———. *The Logic of the Sciences and the Humanities.* New York: Macmillan, 1947.

Ravindran, Ravi. "Perception in Physics and Yoga." *ReVision* 3 (1980): 36–43.

Restivo, Sal P. "Parallels and Paradoxes in Modern Physics and Eastern Mysticism." *Social Studies of Science* 8 (1978): 143–81; 12 (1982): 37–71.

de Riencourt, Amaury. *The Eye of Shiva: Eastern Mysticism and Science.* New York: William Morrow and Company, 1981.

Roszak, Theodore. "The Monster and the Titan: Science, Knowledge, and Gnosis." *Daedalus* (Summer 1974), pp. 17–32.

Sachs, Mendel. "Comparison of the Field Concept of Matter in Relativity Physics and the Buddhist Idea of Nonself." *Philosophy East and West* 33 (1983): 395–99.

Siu, R. G. H. *The Tao of Science: An Essay on Western Knowledge and Eastern Wisdom.* Cambridge: MIT Press, 1957.

Schrödinger, Erwin, *My View of the World.* Cambridge: Cambridge University Press, 1964.

Smith, Huston. *Forgotten Truth.* New York: Harper and Row, 1976.

Synder, Paul. *Toward One Science.* New York: St. Martin's Press, 1978.

Talbot, William. *Mysticism and the New Physics.* New York: Bantam Books, 1980.

Thomsen, Dietrick E. "Mystic Physics." *Science News* 116 (August 4, 1979): 94.

Wilbur, Ken. "Physics, Mysticism and the New Holographic Paradigm: A Critical Appraisal" and "Reflections on the New-Age Paradigm: A Conversation with Ken Wilbur." In *The Holographic Paradigm and Other Paradoxes,* edited by Ken Wilbur. Boulder: Shambhala Publications, 1982.

Yukawa, Hideki. *Creativity and Intuition: A Physicist Looks at East and West.* New York: Kodansha International Ltd., 1973.

Zukav, Gary. *The Dancing Wu Li Masters.* New York: William Morrow and Company, 1979.

2. Science and Western Religion

Barbour, Ian G. *Issues in Science and Religion.* New York: Harper and Row, 1966.

———. *Myths, Models and Paradigms.* New York: Harper and Row, 1974.

———., ed. *Science and Religion: New Perspectives on the Dialogue.* New York: Harper and Row, 1968.

Bedau, Hugo Adam. "Complementarity and the Relation Between Science and Religion." *Zygon* 9 (1974): 202–24.

Benjamin, A. Cornelius. "Mysticism and Scientific Discovery." *Journal of Religion* 36 (1956): 169–76.

Davies, Paul. *God and the New Physics.* New York: Simon and Schuster, 1983.

Ferré, Frederick. "Metaphors, Models and Religion." *Soundings* 51 (1968): 327–45.

———. "Mapping the Logic of Models in Science and Theology." In *New Essays on Religious Language,* edited by Dallas M. High. New York: Oxford University Press, 1969.

———. "Explanation in Science and Religion." In *Earth Might Be Fair,* edited by Ian G. Barbour. Englewood Cliffs: Prentice-Hall, 1972. Pp. 14–32.

Heisenberg, Werner. *Across the Horizons.* New York: Harper and Row, 1974.

———. *Physics and Beyond.* New York: Harper and Row, 1971.

Peacocke, A. R., ed. *The Sciences and Theology in the Twentieth Century.* South Bend: University of Notre Dame Press, 1981.

Ramsey, Ian T., *Religion and Science: Conflict and Synthesis.* London: S.P.C.K., 1964.

Schilling, Harold K. *The New Consciousness in Science and Religion.* Philadelphia: United Church Press, 1973.

Townes, Charles H. "The Convergence of Science and Religion." *Zygon* 1 (1966): 301–11.

Wiebe, Don. "Science and Religion: Is Compatibility Possible?" *Journal of American Scientific Affiliation* 30 (1978): 169–78.

3. Philosophy of Science

Agassi, Joseph. "The Nature of Scientific Problems and Their Roots in Metaphysics." In *The Critical Approach to Science and Philosophy,* edited by Mario Bunge. New York: The Free Press, 1964.

Broad, William and Nicholas Wade. *Betrayers of the Truth: Fraud and Deceit in the Halls of Science.* New York: Simon and Schuster, 1982.

Brown, Harold I. *Perception, Theory and Commitment: The New Philosophy of Science.* Chicago: University of Chicago Press, 1979.

Bunge, Mario. *Intuition and Science.* Englewood Cliffs: Prentice-Hall, 1962.

———. "The Weight of Simplicity in the Construction and Assaying of Scientific Theories." In *Probability, Confirmation and Simplicity,* edited by M. Foster and M. Martin. New York: Odyssey Press, 1966, pp. 280–309.

Burtt, Edwin A. *The Metaphysical Foundations of Modern Physical Science.* Garden City: Doubleday, 1964.

Carnap, Rudolf. *An Introduction to the Philosophy of Science.* Edited by Martin Gardner. New York: Basic Books, 1966.

Danto, Arthur C. and Sidney Morgenbesser, eds. *Philosophy of Science.* New York: New American Library, 1960.

Duhem, Pierre. *The Aim and Structure of Physical Theory.* Translated by Philip P. Weiner. New York: Atheneum, 1974.

Eddington, Sir Arthur. *The Nature of the Physical World.* Ann Arbor: University of Michigan Press, 1958.

Feyerabend, Paul. *Against Method: An Outline of an Anarchistic Theory of Knowledge.* London: NLB, 1975.

———. "Explanation, Reduction, and Empiricism." In *Scientific Explanation, Space, and Time,* edited by Herbert Feigl and Grover Maxwell. Minneapolis: University of Minnesota Press, 1962, pp. 28–97.

———. *Science in a Free Society.* London: NLB, 1978.

Hacking, Ian. *Representing and Intervening: Introductory Topics in the Philosophy of Natural Science.* Cambridge: Cambridge University Press, 1983.

Hanson, Norwood Russell. *Perception and Discovery: An Introduction to Scientific Inquiry.* Edited by Willard C. Humphreys. San Francisco: Freeman, Cooper & Co., 1969.

———. *Observation and Explanation: A Guide to Philosophy of Science.* New York: Harper and Row, 1971.

———. *Patterns of Discovery: An Inquiry into the Conceptual Foundations of Science.* Cambridge: Cambridge University Press, 1972.

Heisenberg, Werner. *Physics and Philosophy.* New York: Harper and Row, 1958.

———. "The Representation of Nature in Contemporary Physics." In *Symbolism in Religion and Literature,* edited by Rollo May. New York: George Braziller, 1960.

Hempel, Carl G., *Aspects of Scientific Explanation and Other Essays in Philosophy of Science.* New York: Free Press, 1965.

———. *Philosophy of Natural Science.* Englewood Cliffs: Prentice-Hall, 1966.

Hesse, Mary B. *Models and Analogies in Science.* South Bend: University of Notre Dame Press, 1970.

———. *The Structures of Scientific Inference.* Berkeley/Los Angeles: University of California Press, 1974.

Kuhn, Thomas S. *The Essential Tension: Selected Studies in Scientific Tradition and Change.* Chicago: The University of Chciago Press, 1977.

———, *The Structures of Scientific Revolutions.* 2d ed. Chicago: The University of Chicago Press, 1970.

Lakatos, Imre, and Alan Musgrave, eds. *Criticism and the Growth of Knowledge.* Cambridge: Cambridge University Press, 1970.

Nagel, Ernest. *The Structure of Science: Problems in the Logic of Scientific Explanation.* New York: Harcourt, Brace and World, 1961.

———. "The Nature and Aim of Science." In *Philosophy of Science Today,* edited by Sydney Morgenbesser. New York: Basic Books, 1967.

Poincaré, Henri. *Science and Hypothesis.* New York: Dover, 1952.

———. *The Value of Science.* New York: Science Press, 1907.

Polanyi, Michael. *Personal Knowledge.* New York: Harper and Row, 1958.

Popper, Karl. *The Logic of Scientific Discovery.* New York: Basic Books, 1959.

Quine, Willard Van Orman. *From a Logical Point of View.* New York: Harper and Row, 1961.

———. *The Ways of Paradox and Other Essays.* New York: Random House, 1966.

Scheffler, Israel. *Science and Subjectivity.* New York: Bobbs-Merrill, 1967.

Schilipp, Paul Arthur, ed. *Albert Einstein: Philosopher-Scientist.* LaSalle, Ill.: Open Court, 1949.

Shapere, Dudley. "The Structure of Scientific Revolution." *Philosophical Review* 73 (1964): 383–94.

———. "Meaning and Scientific Change." In *Mind and Cosmos,* edited by Robert Colodny. Pittsburgh: University of Pittsburgh Press, 1966, pp. 41–85.

Suppe, Frederick, ed. *The Structure of Scientific Theories.* 2d ed. Urbana: University of Illinois Press, 1977.

Toulmin, Stephen. *The Philosophy of Science: An Introduction.* New York: Harper and Row, 1953.

———. "Contemporary Scientific Mythology." In *Metaphysical Beliefs,* edited by Stephen Toulmin et al. New York: Schocken Books, 1957.

———. *Foresight and Understanding.* New York: Harper and Row, 1961.

———. "Conceptual Revolutions in Science." *Synthese* 17 (1967): 75–91.

———. "Philosophy of Science." *Encyclopedia Britannica.* 15th ed. 16: 375–93.

Watkins, J. W. N. "Confirmable and Influential Metaphysics." *Mind* 67 (1958): 344–65.

Wisdom, J. O. "Scientific Theory: Empirical Content, Embedded Ontology, and Weltanschauung." *Philosophy and Phenomenological Research* 33 (1972): 62–76.

4. Studies of Mysticism

Alston, William. "Ineffability." *Philosophical Review* 65 (1956): 506–22.

Bharati, Agehananda. *The Light at the Center.* Santa Barbara: Ross-Erickson, 1976.

Danto, Arthur C. "Language and the Tao: Some Reflections on Ineffability." *Journal of Chinese Philosophy* 1 (1973): 45–55.

Eliade, Mircea. *Shamanism.* Princeton: Princeton University Press, 1964.

Happold, F. C. *Mysticism: A Study and an Anthology.* Baltimore: Penguin Books, 1971.

Henle, Paul. "Mysticism and Semantics." *Philosophy and Phenomenological Research* 9 (1949): 416–22.

Huxley, Aldous. *The Perennial Philosophy.* New York: Harper and Row, 1944.

James, William. *The Varieties of Religious Experiences.* New York: New American Library, 1958.

Jones, Richard H. "A Philosophical Analysis of Mystical Utterances." *Philosophy East and West* 29 (1979): 255–74.

———. "Experience and Conceptualization in Mystical Knowledge." *Zygon* 18 (1983): 139–65.

———. "Must Enlightened Mystics Be Moral?" *Philosophy East and West* 34 (1984): 273–93.

Katz, Steven T., ed. *Mysticism and Philosophical Analysis.* New York: Oxford University Press, 1978.

Otto, Rudolf. *Mysticism East and West.* New York: Macmillan, 1957.

Perovich, Jr., Anthony N. "Mysticism and the Philosophy of Science." *Journal of Religion* 65 (1985): 63–82.

Pletcher, Galen K. "Agreement Among the Mystics." *Sophia* 11 (1972): 5–15.

———. "Mysticism, Contradiction, and Ineffability." *American Philosophical Quarterly* 10 (1973): 201–11.

Smart, Ninian. "Interpretation and Mystical Experience." *Religious Studies* 1 (1965): 75–87.

Staal, J. Frits. *Exploring Mysticism.* Berkeley/Los Angeles: University of California Press, 1975.

Stace, Walter. *Mysticism and Philosophy.* Philadelphia: J. B. Lippincott, 1960.

Tillich, Paul. *Dynamics of Faith.* New York: Harper and Row, 1957.

———, "The Nature of Religious Language." In his *Theology of Culture,* edited by Robert C. Kimball. New York: Oxford University Press, 1959.

Underhill, Eveyln. *Mysticism.* New York: E. P. Dutton, 1961.

———. *Practical Mysticism.* New York: E. P. Dutton, 1915.

Wainwright, William J. "Mysticism and Sense Perception." *Religious Studies* 9 (1973): 257–78.

———. *Mysticism.* Madison: University of Wisconsin Press, 1981.

Woods, Richard, O.P., ed. *Understanding Mysticism.* Garden City: Doubleday, 1980.

Zaehner, R. C. *Mysticism: Sacred and Profane.* New York: Oxford University Press, 1961.

5. Secondary Literature on Indian Religious Traditions

Basham, A. L. *The Wonder That Was India.* New York: Grove Press, 1954.

Conze, Edward. *Buddhist Meditation.* New York: Harper and Row, 1956.

———. *Buddhist Thought in India.* London: George Allen and Unwin, 1962.

Coomaraswamy, A. *Time and Eternity.* Ascona, Switzerland: Artibus Asiae, 1947.

Danto, Arthur C. *Mysticism and Morality.* New York: Harper and Row, 1972.

Dasgupta, Surendranath. *A History of Indian Philosophy.* 5 vols. Cambridge: Cambridge University Press, 1922–1955.

Deutsch, Eliot. *Advaita Vedanta: A Philosophical Reconstruction.* Honolulu: The University Press of Hawaii, 1969.

Dutt,. Nalinaksha. *Early Monastic Buddhism.* 2d ed. Calcutta: Firma K. L. Mukhopadhyay, 1971.

Edgerton, Franklin. "The Upaniṣads: What Do They Seek and Why?" *Journal of the American Oriental Society* 49 (1929): 97–121.

———, "Did the Buddha Have a System of Metaphysics?" *Journal of the American Oriental Society* 79 (1959): 81–85.

Eliade, Mircea. "Indian Symbolisms of Time and Eternity." In his *Images and Symbols,* translated by Philip Mariet. New York: Sheed and Ward, 1969, Pp. 57–91.

———. *Yoga: Immortality and Freedom.* Princeton: Princeton University Press, 1969.

———. *Patanjali and Yoga.* New York: Schocken, 1975.

Gombrich, R. F. "Ancient Indian Cosmology." In *Ancient Cosmologies,* edited by Carmen Blacker and Michael Loewe. London: George Allen and Unwin, 1975. Pp. 110–42.

Ingalls, Daniel H. H. "Śaṁkara on the Question: Whose is Avidyā?" *Philosophy East and West* 3 (1953): 69–72.

———. "Śaṁkara's Arguments Against the Buddhists." *Philosophy East and West* 3 (1954): 291–306.

Jayatilleke, K. N. *Early Buddhist Theory of Knowledge.* London: George Allen and Unwin, 1963.

Johansson, Rune. *The Psychology of Nirvana.* London: George Allen and Unwin, 1969.

Jones, Richard H. "Jung and Eastern Religious Tradition." *Religion* 9 (1979): 141–56.

———. "The Nature and Function of Nāgārjuna's Arguments." *Philosophy East and West* 28 (1978): 485–502.

———. "Theravāda Buddhism and Morality." *Journal of the American Academy of Religion* 47 (1979): 371–87.

———. "*Vidyā* and *Avidyā* in the *Īśa Upaniṣad.*" *Philosophy East and West* 31 (1981): 79–87.

Kalupahana, David J. *Causality: The Central Philosophy of Buddhism.* Honolulu: The University Press of Hawaii, 1975.

———. *Buddhist Philosophy: A Historical Analysis.* Honolulu: The University Press of Hawaii, 1976.

Karunadasa, Y. *Buddhist Analysis of Matter.* Colombo: Department of Cultural Affairs, 1967.

King, Winston L. *Theravāda Meditation: The Buddhist Transformation of Yoga.* University Park: The Pennsylvania State University Press, 1980.

Mahadeva, T. M. P. *Gauḍapadā: A Study in Early Advaita.* Madras: University of Madras, 1954.

Mitchell, Donald W. "Analysis in Theravāda Buddhism." *Philosophy East and West* 21 (1971): 23–31.

Ñāṇananda, Bhikkhu. *Concept and Reality in Early Buddhist Thought.* Kandy: Buddhist Publication Society, 1971.

Nyanaponika, Bhikkhu. *The Heart of Buddhist Meditation.* New York: Samuel Weiser, 1973.

Potter, Karl. *Presuppositions of India's Philosophies.* Englewood Cliffs: Prentice-Hall, 1963.

———, ed. *Advaita Vedānta up to Śaṃkara and His Pupils.* Princeton: Princeton University Press, 1981.

Rahula, Walpola. *What the Buddha Taught.* New York: Grove Press, 1959.

Robinson, Richard. *The Buddhist Religion.* Belmont: Dickenson, 1970.

Smart, Ninian. *Doctrine and Argument in Indian Philosophy.* London: George Allen and Unwin, 1964.

Staal, J. F. *Advaita and Neo-Platonism: A Critical Study in Comparative Philosophy.* Madras: University of Madras, 1961.

Swearer, Donald K. "Two Types of Saving Knowledge in the Pali Suttas." *Philosophy East and West* 22 (1972): 355–71.

Underwood, Frederic B. "Buddhist Insight: The Nature and Function of *Paññā* in the Pāli Nikāyas." Ph.D. diss., Columbia University, 1973.

Wayman, Alex, "The Meaning of Unwisdom *(Avidyā).*" *Philosophy East and West* 7 (1957): 21–25.

———. "Buddhist Dependent Origination." *History of Religions* 10 (1971): 185–203.

Williams, David M. "The Translation and Interpretation of the Twelve Terms in the *Paṭiccasamuppāda.*" *Numen* 21 (1974): 35–63.

Zaehner, Robert C. *Hinduism.* New York: Oxford University Press, 1962.

———. *Hindu and Muslim Mysticism.* New York: Schocken, 1969.

Index

Abhidhamma, 53, 86–87, 143, 186–87, 239
Adhyāsa (superimposition), 64, 148, 152
Alston, William, 225
Apavāda (desuperimposition), 69, 152
Austin, William H., 247
Avataṁsaka Buddhism, 189–92, 196, 206
Avyākata (unanswered questions), 49, 55–56, 142
Ayer, A. J., 135, 145, 223, 251

Babylonian astronomy, 23
Bacon, Roger, 19, 123–24
Barbour, Ian, 90, 171–72, 194, 243, 246
Beauregard, Olivier Costa de, 239
Being: and becoming, 43–45, 91–92, 93–94, 176, 186–91, 203–4, 214–18; and structure, 91–92, 185–86, 214–18
Belief, 12, 41–42, 57
Bellah, Robert N., 252
Benson, Herbert, 219, 221, 223, 227
Berger, Peter, 77, 158, 236, 238, 244, 252
Bernstein, Jeremy, 246
Bhāvanā (meditation), 53–54, 88, 251
Biofeedback, 252
Black, Max, 245
Bohm, David, 147, 191, 247, 248, 249
Bohr, Niels, 38, 155, 158, 176, 248
Bondi, Hermann, 234
Boyle, Robert, 75
Brahman/*ātman* (self), 60–63, 64–66, 105–7
Brahmavidyā. See *Jñāna*
Broad, C. D., 128
Broad, William, 237–38, 242
Buber, Martin, 127
Buddhaghosa, 86
Bunge, Mario, 139, 235
Burhoe, Ralph W., 246
Burtt, Edwin, 123, 240

Čapek, Milač, 239
Capra, Fritjof, 11, 170, 174, 187, 201–4, 246
Carnap, Rudolf, 240
Cartwright, Nancy, 235
Causation, 99–101, 105–6
Chang, Garma, C. C., 189
Chari, C. T. K., 249
Chew, Geoffrey, 171, 190, 202
Chuang tzu, 13
Chu Hsi, 211
Clifford, W. K., 250
Complementarity, 174, 176–78, 199, 205–6
Concepts and conceptualizations, 19–20, 44, 53, 55, 137–38, 143–45, 147–50, 166, 170, 198–99. *See also* Language
Conditionality, 188–91
Consciousness, 21; physiological accounts of, 219–22; states of, 127–30
Conventionalism, 36
Conze, Edward, 236, 237
Cooper, Leon, 246
Copenhagen Interpretation of quantum physics, 154–55, 176
Copernican revolution, 19, 26, 27, 29, 37, 78, 183
Cosmology, 73–74, 180–84, 187

Danto, Arthur, 146, 236, 245
Darwin, Charles, 23, 78
Darwinian theory, 19, 32
Davies, Paul, 247
Deikman, Arthur J., 197, 239
Descartes, René, 67, 116
Deutsch, Eliot, 103
Dewey, John, 89
Dhammā (elements of the experienced world), 52–53, 56, 86–87, 104, 143, 149, 186–87
Dōgen, 46, 213
Douglas, Mary, 77
Duhem, Pierre, 24, 91, 235
Dukkha (suffering), 48–50
Duncan, Ronald, 125
Durkheim, Émile, 228, 252

Eckhart, Meister, 178, 199
Eddington, Arthur, 36, 74, 226
Edgerton, Franklin, 237
Einstein, Albert, 19, 75, 97, 124, 128, 179, 184, 231. *See also* Relativity
Eliade, Mircea, 239, 252
Evans-Pritchard, E. E., 228
Experience. *See* Mystical experiences; Observation
Explanation, 23

Facts, 22–23, 25–26, 29–30, 39, 78, 84, 212
Feinberg, Gerald, 236
Fenton, John Y., 253
Feyerabend, Paul, 31–32, 115, 235
Fromm, Erich, 238

Galileo, 22, 33, 147, 246
Geertz, Clifford, 115, 236
Gellner, Ernest, 89
Gödel, Kurt, 239
Gombrich, R. F., 182
Goodenough, Erwin, 125, 246
Goodman, Nelson, 236
Govinda, Lama Anagarika, 239
Greene, John, 253
Grimm, George, 57–58

Haack, Susan, 245
Hamnett, Ian, 253
Hanson, Norwood Russell, 23, 26, 147, 235, 240, 245, 250
Heisenberg, Werner, 13, 24, 146, 159, 166, 172, 176, 185, 186, 193, 212, 246
Heisenberg Uncertainty Principle, 154, 193, 195
Hempel, Carl, 24, 25, 159, 234
Henle, Paul, 245
Henry, Granville C., Jr., 246
Heraclitus, 206
Hesse, Mary, 33, 157, 158–59, 235
Holisms, 188–91, 192–94
Holography, 191–92
Holton, Gerald, 236, 246
Hooykaas, R., 238
Horton, Robin, 77
Huxley, Aldous, 128

Idealism, 36–37
Inada, Kenneth, 105
Ineffability, 150–53
Ingalls, Daniel H. H., 237
Insight experiences, 115, 130–32
Instrumentalism, 37
Ionian astronomy, 23
Īśvara, 48, 63

Jaina, Padmanabh S., 241
James, William, 88, 125, 128, 226, 250
Jastrow, Robert, 74, 180
Jayatilleke, K. N., 87, 118–23, 143, 182, 241, 244
Jhānā/dhyānas (trances), 48, 49, 54, 68, 119–20, 129–30, 224
Jñāna (knowledge of Brahman), 66–70, 111–12, 116–17
Jung, Carl, 101, 246

Kalupahana, David, 118–23, 241
Karma/kamma, 50, 95, 102–5, 118–23, 164, 176, 183, 239
Karunadasa, Y., 240
Kepler, Johannes, 23, 80
Kierkegaard, Sören, 213
King, Winston L., 182, 237, 241, 245
Kline, Morris, 245
Kluckhohn, Clyde, 77
Koans, 204–5
Kuhn, Thomas, 31, 32, 235

Lakatos, Imre, 28, 75, 117, 164, 235
Langer, Suzanne K., 235
Language, 43, 65, 140–65; mirror theory of, 146–50, 153, 155–56, 161–62, 165, 217
La Vallee Poussin, Louis de, 182
Laws, 21, 37, 100–101
Lawson, E. Thomas, 252
Leach, Edmund, 228, 229
Lee, D. D., 244
Leibniz, G. W. V., 27, 97, 152
LeShan, Lawrence, 172
Lessa, William A., 236
Levels of organization, 37–40, 194–97, 216–17, 244
Levels of truth, 244
Little, David, 243
Luckmann, Thomas, 144, 158, 238, 244
Lysenko, T. D., 75

McCauley, Robert N., 252
Mach, Ernst, 125, 190
MacIntyre, Alasdair, 253
Mādhyamika Buddhism, 59, 186–87, 189
Mahātissa, 86, 137
Malcolm, Norman, 251–52
Malinowski, Bronislaw, 238
Maslow, Abraham, 125, 231
Mathematics, 164–66
Māyā (creative power), 63–64, 143, 162, 178, 201
Meaningfulness of utterances, 140–43
Merton, Robert K., 237
Metaphoric utterances, 156–64
Metaphysics, 33–35, 81–92, 115, 160, 225
Michelson-Morley experiment, 29, 235
Mind and body, 62–63, 83, 192–94, 205–6, 219, 238, 241
Mindfulness, 43–46, 51, 53–54, 69, 86–88, 137–39, 179, 216. *See also* Observation
Minkowski, Hermann, 206
Models, 24–25, 32, 159–60
Moggalāna, 103
Mokṣa (liberation), 66, 69–70
Morgenbesser, Sidney, 89
Musila, 45
Mystery, 123–26, 215
Mystical experiences, 42–47, 126, 127; in con-

text, 47–48, 113–18; physiological accounts of, 46–47, 129–30, 222–27
Mystical knowledge, 45–47, 48, 110–18
Mystical oneness, 185–86

Nāgārjuna, 189
Nagel, Ernest, 234, 235
Ñaṇanda, 244
Narada, 45
Naranjo, Claudio, 236
Needleman, Jacob, 213, 243
Newton, Isaac, 21, 23, 24, 75, 80, 128, 130, 150, 231
Newtonian physics, 22, 27, 28, 30, 32, 36, 81, 93, 97, 158
Nibbāna, 54–56, 57–58, 87, 95, 151, 222, 242
Northrop, F. S. C., 177, 197–200
Nyanaponika, Thera, 239

Objectivity, 21, 132–36, 137–39
Observation, 20–21, 25, 43, 136–39
Oppenheimer, J. Robert, 176, 196, 249
Ornstein, Robert, 221, 235, 236, 239

Paññā (insight), 51, 54
Papañca (conceptual projection), 55, 142, 148
Paradox, 153–56, 204–5
Parāvidyā. See *Jñāna*
Park, James L., 247
Parmenides, 43
Paṭicca-samuppāda (dependent origination), 50–51, 104–5
Pauli, Wolfgang, 101, 146
Peirce, C. S., 132
Pelletier, Kenneth R., 169, 196, 249, 252
Penfield, Wilder, 252
Penner, Hans H., 252–53
Penrose, R., 247
Pepper, Stephen, 238
Planck, Max, 31
Plato, 160–61
Plotinus, 199
Poincaré, Henri, 36, 235, 239
Popper, Karl, 25, 29, 243, 244
Potter, Karl, 240
Prajñāpāramitā Buddhism, 187, 217
Psychology, 77, 129–30, 169, 222, 224, 227–33; Gestalt images, 21, 28, 29, 46
Ptolemaic astronomy, 25, 27, 32, 135
Pūrva Mīmāṃsā, 143

Quine, Willard Van Orman, 25, 146, 165, 234

Rāmānuja, 60
Ravetz, Jerome, 234, 244
Realism, 37, 56, 174
Reality, 20, 35–37, 39–41, 57–58, 64–66, 88–90
Rebirth, 49–50, 118–23
Reductionism, 38–39, 228–31
Relativity, 30, 36, 93, 97, 184–85
Religion, 41–42, 48–49, 75, 78–79, 82, 160; and science, 174–76; socioscientific studies of, 227–33
Ṛta (order), 101
Rudner, Richard, 238
Russell, Bertrand, 125, 137
Ryle, Gilbert, 236

Samādhi (concentration), 51, 54, 68, 224
Sāṁkhya-Yoga, 47, 63, 105, 114, 171, 199, 201, 247, 249
Sammādiṭṭhī (right views), 51–52, 85–86, 148–49
Saṅkhārā (constructed entities), 49, 52
Scheffler, Israel, 243, 245
Schilling, Harold K., 242, 245
Schrödinger, Erwin, 80, 170, 192, 196
Self, conceptions of, 43, 44, 47, 52–54, 62–63, 64, 67, 86, 115–16, 142, 217. *See also* Brahman/*ātman*
Siddhis/*iddhī* (paranormal powers), 95, 107–9, 118, 169, 227, 248
Simon, Herbert A., 243
Siu, R. G. H., 170
Smart, Ninian, 103, 116, 117, 121, 236
Smith, Huston, 78–79
Space, 52, 97–99
Spiro, Melford, 229, 239
Staal, J. F., 85, 114, 116–17, 252
Stevenson, Ian, 242
Strawson, Peter F., 85
Suññatā/*śūnyatā* (voidness), 52, 186–88, 189, 198
Suzuki, D. T., 189, 190, 191, 246
Symbolism, 114, 158, 160–62

Takka (reasoning), 113, 165
Tantric Buddhism, 204
Taoism, 11, 177–78
Tart, Charles, 62, 129, 236, 251
Taylor, Paul W., 242
Teilhard de Chardin, Pierre, 75
Tennant, F. R., 114
Theory, 21–23, 90, 117–18, 121–23, 141; theoretical change, 26–28; theory-acceptance, 28–33
Tillich, Paul, 161–62
Time, 92–97, 195
Toulmin, Stephen, 19, 23, 29, 37, 159, 202, 234, 237, 253
Transcendental Meditation, 223
Twiss, Sumner B., Jr., 243

Underhill, Evelyn, 253

Vedas, 59, 67, 82, 101, 114, 117, 143, 181, 224
Volt, Evon Z., 236

Wade, Nicholas, 237–38, 243
Wainwright, William J., 252
Watkins, J. W. N., 239
Watts, Alan, 196, 236
Wayman, Alex, 239
Ways of life, 41–42, 73–76
Weinberg, Steven, 73, 235
Welbon, Guy, 236
Weston-Smith, Miranda, 125
Weyl, Hermann, 93
Whitehead, Alfred North, 136
Wigner, Eugene P., 246, 249
Wisdom, J. O., 238–39
Wittgenstein, Ludwig, 150, 232
World-views, 52, 77–78, 114, 118, 201, 203

Yoga, 68, 247
Yogācāra Buddhism, 59, 108
Yonan, Edward A., 252–53

Zen Buddhism, 198, 204–5
Zukav, Gary, 11, 171, 195, 204–6, 246